Assessing Bioavailability *of* Drug Delivery Systems

Mathematical Modeling

Assessing Bioavailability *of* Drug Delivery Systems

Mathematical Modeling

Jean-Maurice Vergnaud
Iosif-Daniel Rosca

Taylor & Francis
Taylor & Francis Group

Boca Raton   London   New York   Singapore

A CRC title, part of the Taylor & Francis imprint, a member of the
Taylor & Francis Group, the academic division of T&F Informa plc.

Published in 2005 by
CRC Press
Taylor & Francis Group
6000 Broken Sound Parkway NW, Suite 300
Boca Raton, FL 33487-2742

© 2005 by Taylor & Francis Group, LLC
CRC Press is an imprint of Taylor & Francis Group

No claim to original U.S. Government works
Printed in the United States of America on acid-free paper
10 9 8 7 6 5 4 3 2 1

International Standard Book Number-10: 0-8493-3044-0 (Hardcover)
International Standard Book Number-13: 978-0-8493-3044-5 (Hardcover)

Library of Congress Cataloging-in-Publication Data

Catalog record is available from the Library of Congress

Taylor & Francis Group
is the Academic Division of T&F Informa plc.

Visit the Taylor & Francis Web site at
http://www.taylorandfrancis.com

and the CRC Press Web site at
http://www.crcpress.com

Preface

This book intends to assist the following four important categories of people involved in curing patients:

- Researchers at university and in the pharmaceutical industry in charge of making discoveries of new drugs or, perhaps much better, improved dosage forms
- Therapists devoted to curing patients at the hospital
- Physicians and pharmacists permanently in touch with their ill customers, who must give these patients advice
- Last but not least, professors—and more especially their students, who will be the physicians, pharmacists, and researchers of tomorrow

Researchers in the pharmaceutical industry know that modern drug discovery, as well as the dosage forms through which the drug should be administered, has become an increasingly time-consuming and expensive process. Extensive testing in clinical trials follows a long laboratory research and development phase. In the end, the development of a single drug can be the result of more than 10 to 15 years of work. Researchers must create a compound that delivers itself to the appropriate target with low toxicity and few side effects.

Clinical testing has three main phases:

- Phase 1, conducted on healthy volunteers, enables a drug's safety, identifies side effects, and determines appropriate dosages.
- Phase 2 is conducted on a larger group of patients to further evaluate safety and to generate preliminary data on effectiveness.
- Phase 3, consisting of more than 1000 patients, confirms effectiveness, creates a side-effect profile, and compares the effects of the drug to other treatments.

During these three phases, when effectiveness of the dosage form should be demonstrated and side effects at least reduced, the knowledge provided by this book can be useful in two ways: reducing the time of these studies by offering shortcuts and a kind of Ariadne's thread to researchers as far as the dosage form is concerned, and providing the basic knowledge necessary to improve the main qualities of the dosage form.

In every phase, it would be of great interest to test oral sustained-release dosage forms. As a matter of fact, the accuracy attained in evaluating the plasma drug profile obtained with these dosage forms is much better than for their conventional counterparts; because the experimental time is longer, the main characteristics, such as the time at the peak and the half-life time, are measured more easily. Moreover, by using the methods described in the book, it becomes easier to select the right times at which to take samples.

Therapists at hospitals and health clinics, medical doctors, and pharmacists must deal with patients in order to adjust the drug delivery schedule with the right dosage over the right period of time. New technology can open up additional pricing options and expand a product line, and value comes not only from effectiveness but also from improved patient compliance. Patients and therapists find value in reducing side effects and simplifying dosing, improving the quality of life and ease of use. On the one hand, Chapter 2, devoted to intravenous drug delivery through either repeated injections or continuous infusion, would be of interest to these professionals. Because all patients, depending on their age or type of illness (among other things), exhibit a wide range of responses to a drug, a kind of pharmacokinetic profile should be defined for each patient. The purpose of Chapter 2 is to help the therapists to evaluate the typical pharmacokinetic parameter, and by doing so, to adapt the correct regimen to each patient. A method is described that allows therapists to evaluate these parameters at the very beginning of the treatment. On the other hand, oral controlled release has been a big winner for drug delivery. In fact, this oral delivery device, taken once-a-day, is not only a commodity item, but also the right way to adjust the right dose over a long period of time, reducing the difference in the levels of the peaks and troughs and thereby reducing harmful side effects.

Physicians and pharmacists who are permanently in touch with the patients must teach those patients to realize the two main facts concerning the oral dosage forms with controlled release. First, these therapeutic systems offer important advantages over immediate-release dosages for diseases that require the most constant possible effective blood levels over prolonged durations of therapy—advantages including a decrease in the number and frequency of side effects—without having to place more importance upon the easier compliance brought by the once-a-day dosing that is so easy to consent to. Second, regarding the issue of compliance, it is important that patients understand that omission in missing a dose is one mistake, but trying to compensate for this oversight by doubling the dose the following day is another, much worse mistake, called *wrong dosage*. The drug concentration profiles drawn in the various figures collected in Chapter 8 clearly show that impermissible delay, followed by execution at the wrong time in the wrong mode, may be a source of harmful problems. This chapter also shows, as far as compliance is concerned, that oral dosage forms with controlled release are of great benefit to the patient.

Because we are university professors working either for pharmaceutical industries or for hospitals, we are well aware of the importance of modern university instruction. The university students of today are the therapists and researchers of the future. So, they must understand the importance of drug delivery and the essential part played by oral controlled-release dosage forms, as well as repeated intravenous injections. These students must consider the fact that different patients behave quite differently toward a drug, and that the best way to cure a patient is to adapt the therapy conditions to that patient's own pharmacokinetic parameters, without forgetting the pharmacodynamics. From one patient to another, the main pharmacokinetic parameter may vary greatly within a wide range (e.g., the apparent plasmatic volume from 1 to 3, the half-life time from 1 to 4). A method is described as clearly as possible, using a mathematical treatment, to make the determination of these parameters possible at the very beginning of the therapy. This method also allows students to adapt the appropriate therapy conditions. This method is applied essentially to intravenous delivery but also to oral conventional dosage forms.

Because the process of drug transport with oral dosage forms with sustained release is far more complex than for other methods of drug delivery, whether they are controlled by diffusion, by erosion, or by both, mathematical treatment is impractical and numerical models have been built. Special attention has been paid to these sustained-release dosage forms; not only do they help with patient compliance, but they also reduce side effects and thus participate effectively in modern efficient therapy. Finally, master curves are built especially for intravenous delivery, and we encourage readers to use them, because these can be applied to various drugs when their pharmacokinetic parameters are known. In the same way, a dimensionless number is used for the oral dosage forms by considering the amount of drug delivered to the blood as a fraction of the amount of drug initially in the dosage form; however, the plasma drug concentration is also used, as well as the way to determine it from the value of this previously mentioned dimensionless number.

Authors

Jean-Maurice Vergnaud has retired from teaching, but not from research. He is head of the Laboratory of Materials and Chemical Engineering at the University of Saint-Etienne in Saint-Etienne, France and the author of various books derived from his research. These include *Liquid Transport Processes in Polymeric Materials: Modelling and Industrial Applications*, Prentice Hall, Englewood Cliffs, NJ, 1991; *Drying of Polymeric and Solid Materials: Modelling and Industrial Applications*, Springer-Verlag, New York, 1992; *Cure of Thermosetting Resin: Modelling and Experiments* (with J. Bouzon), Springer-Verlag, New York, 1992; and *Controlled Drug Release of Oral Dosage Forms*, Ellis Horwood Series in Pharmaceutical Technology, New York, 1993.

Iosif-Daniel Rosca is an associate professor in the Department of Polymer Chemistry and Technology at Polytechnic University in Bucharest, Romania.

Acknowledgments

Many colleagues and students who have supported our efforts and brought contributions are worth noting.

Prof. J.M. Vergnaud

Deep gratitude is extended to my colleague M. Rollet, who introduced me to the world of Galenic pharmacy, as well as J. Bardon and C. Chomat, from the very beginning, 20 years ago. I am grateful to F. Falson, professor at the University of Lyon, France, who helped me to understand some key problems in pharmacy.

It is a pleasure to give my best thanks to R.W. Jelliffe, professor at the University of Southern California School of Medicine in Los Angeles, who showed me around the world of in vivo calculation with the problems of endocarditis. His valuable advice helped me in writing this book. I appreciate the kind help of Dr. P. Maire, from the Geriatric Hospital of Lyon, as well as the various people of the ADCAPT group. I am much obliged to Dr. J. Paulet, President of the French Order of Pharmacists, and Dr. M. Trouin, pharmacist, who helped us in demonstrating their essential role with their customers, who are also these patients who take therapies at home.

I have appreciated the collaboration of various colleagues: Prof. R. Wise, from the City Hospital of Birmingham, U.K., in drug transport into the blister fluid; Prof. R. Bodmeier and his colleague J. Siepmann, from the Freie Universität of Berlin, in the fundamental approach to the preparation of dosage forms; Prof. D. Breilh and M.C. Saux, from the Hospital Haut-Levêque of Bordeaux, in the drug transport into and through the lungs.

My best appreciation is given to my students who did their best in preparing their theses; there are so many of them that their names cannot be written here.

I cannot forget those people managing international conferences who invited me to deliver plenary lectures, and I am glad to mention among them Prof. M.H. Rubinstein, from the University of Liverpool, and the community members of the U.S. Food and Drug Administration, Washington, D.C., for their estimable appreciation of our methods of in vitro–in vivo calculation.

Contents

1 Definitions

NOMENCLATURE

ABBREVIATIONS AND TERMS

AUC_{0-t} Area under the curve expressing the drug concentration vs. time, from 0 to t.

AUC_{oral}, $AUC_{i.v.}$ Area under the curve obtained with an oral dosage form, with i.v. injection.

GI(T) Gastrointestinal (tract).

i.v. Intravenous.

Dose Amount of drug injected through an i.v. injection.

ED, LD Median effective dose, median lethal dose, respectively.

T.I. Therapeutic index.

SYMBOLS

C Unbound drug concentration in the blood.

C_0, C_∞ Concentration of drug at time 0, at infinite time, respectively.

Cl Clearance (blood volume/time/kg body weight).

Cl_h, Cl_r Hepatic clearance, renal clearance, respectively.

E Ratio of extraction of drug by the liver during the first-pass hepatic.

k_a Rate constant of absorption of the drug (/h).

k_e Rate constant of elimination of the drug (/h).

M_{in} Amount of drug in the dosage form initially in the GI.

M_t Amount of drug remaining in the dosage form at time t.

Q Flow of the blood.

t Time.

t_{max} Time at which the drug concentration in the blood is maximum for oral dosage form.

V_p Apparent volume of distribution, or apparent volume of the blood.

It is useful to define certain important biopharmaceutical concepts at the beginning of a book dealing with the principles of controlled drug delivery and comparing controlled-release oral dosage forms (sustained dosage forms) with the other, conventional forms of administration, such as oral with immediate release or intravenous infusion.

1.1 DRUG (ACTIVE AGENT) AND ITS SUPPLY FORM (DOSAGE FORM)

The definition of a *drug* given by the World Health Organization is "any substance that is used to modify or explore physiological systems or pathological states for the benefit of the recipient." In the scientific literature, the term *drug* is generally used in the sense of a biologically active substance. The drug is the active material that is responsible for a pharmacological effect.

The *dosage form* or the supply form of presentation of the drug is the complete medication. It consists of the active agent and excipients, which are auxiliary substances. In fact, Galenic pharmacy, the science of the formulation of the dosage forms, is a sophisticated technology.

A *proprietary preparation* is a dosage form of defined and constant composition that bears the trademark of the firm that manufactured it.

Excipients, generally biologically inert, are necessary in the dosage form for various reasons: they ensure potency, consistency, and a volume of supply form suitable for patient use.

For immediate-release dosage forms, the excipients commonly used, such as gelatin, lactose, starch, talc, or paraffin, play the role of agent necessary for binding, filling, disintegrating, lubricating, or making more soluble the active agent. They should promptly dissolve either in the gastric or in the intestinal liquid, depending on the nature of the drug.

For sustained-release dosage forms, the excipient plays an important role, because it must control the rate of release of the drug along the gastrointestinal tract. Monolithic devices are obtained by dispersing the drug in an inert polymer acting as a consistent matrix. The polymer must be biocompatible and pass along the gastrointestinal tract (GIT) without any interference with the body. On the whole, biodegradable and nonbiodegradable polymers are the two extreme cases of polymers: the former disintegrate by following a known rate and the latter are stable. With stable polymers, the process of drug release is controlled by diffusion, whereas with biodegradable polymers this process is controlled by both erosion and diffusion.

It is not the purpose of the book to provide an exhaustive list of these polymers. However, some of them, having been well described for many years, are worth noting.

For stable polymers, deep studies were carried out through *in vitro* tests to determine the kinetics of drug release: ethyl cellulose with various plasticizers, whose percentage as additives was considered [1]; polyvinyl acetate and polyvinyl chloride [2]; polyvinyl acetate [3]; Eudragit[R], which denotes a wide family of copolymers of dimethyl aminoethyl acrylate and ethylmethyl acrylate [4, 5]; polyurethane foam [6]; ethylene–vinyl acetate copolymers [7]; 2-hydroxyethyl methacrylate [8].

Various stable polymers for which the process is controlled by diffusion connected with a large swelling extent have also been studied. Among them are the following: (hydroxyethyl) methyl cellulose [9, 10]; polymers of acrylic acid with a high molecular weight, designated by the trademark Carbopol[R] [11]; and copolymers of N-vinyl-pyrrolidone and 2-hydroxyethyl methacrylate [12].

Biodegradable polymers were presented in a large review [13], as well as hydrogels with additives such as sodium alginate or sodium bicarbonate [14, 15], sodium alginate, chitosan HCl, and polyacrylic acid, all three of which exhibit independently some mucoadhesive properties [16].

1.2 BIOPHARMACEUTICS AND ITS GOALS

As defined by Levy in 1958, "Biopharmaceutics is the study of factors that affect the extent and speed of absorption and liberation of the drug from its various physiochemical forms. It is concerned with the dependence of absorption, distribution, metabolism, retention, or excretion of the drug by the patient on the physiochemical characteristics of the dosage form. It deals essentially with the optimum availability of the drug from the dosage form employed for the target site, i.e., the influence of the drug formulation on the biological activity of the drug."

In other words, a pharmacologically active substance is not necessarily an effective remedy; its efficacy depends on the method by which it is delivered to the organism. This fact confers a primary importance on workers dealing with Galenic studies.

The following sections consider stages in succession for the drug in the body.

1.2.1 BIOPHARMACEUTICAL STAGE WITH TWO LINKED STEPS

- Liberation of the drug. Very fast with immediate-release dosage forms, and much slower with sustained-release dosage forms.
- The drug is thus simultaneously dissolved in the GI liquid, either in the stomach or in the intestine, according to the nature of the excipient or the coating.

1.2.2 PHARMACOKINETIC STAGE

This stage has several steps (Figure 1.1):

- Absorption of the dissolved drug into the circulating blood, with a possible metabolism in the liver that reduces the bioavailability of the drug.
- Distribution of the drug that is more or less bound to the plasmatic proteins, leading to an apparent plasmatic volume, which can be much larger than the blood volume.
- Diffusion transport of the drug to the target organ.
- Elimination of the drug, either unchanged or in the form of various metabolites, through the kidney, or via other means of excretion. Elimination is characterized by three pharmacokinetic parameters:
  - Rate of elimination
  - Half-life time
  - Clearance

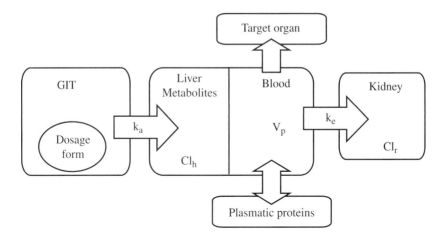

FIGURE 1.1 Path followed by a drug from its release along the gastrointestinal tract to excretion through the kidney.

1.2.3 Pharmacodynamic Stage

During the pharmacokinetic stage, the drug diffuses to the site of action in the target organ, provoking the therapeutic effect.

1.3 PHARMACOKINETICS

Pharmacokinetics is concerned with the kinetics of absorption, distribution, metabolism, and excretion of the drug and its metabolites (if there are any) produced by the body itself. In other words, pharmacokinetics deals with what the body does to the drug.

After the administration of the drug, the following processes occur: release of the drug from the dosage form along the GI tract (liberation); diffusion through the GI membrane into the circulating blood system (absorption); distribution via the circulation and lymphatic system (distribution) to the target organ, where the drug is bound to its receptor and produces its therapeutic action (pharmacodynamics).

Mathematics is a primary tool to study the change in the concentration of the drug (and the metabolites if the data are provided) in the following stages:

- During the stage of drug liberation from the dosage form, especially in the case of sustained release. With immediate release, the total amount of drug is available to cross the GI membrane during the stage of absorption.
- During the stage of absorption into the circulating blood. This process of absorption is assumed to be described by a first-order kinetics with the rate constant of absorption k_a.
- During the stage in which the drug binds to plasmatic proteins, necessitating sometimes the use of the two-compartment model. Binding and unbinding of a drug to plasmatic proteins are each described by first-order kinetics with a rate constant, allowing equilibrium between the

concentrations of bound and unbound drug to be reached promptly. This binding of the drug, when it occurs, reduces the concentration of the free drug in the blood and thus is responsible for an apparent volume of distribution V_d much larger than the volume of the blood itself, because only the concentration of the free drug is measured.

- During the stage of elimination, which is considered to be described by a first-order kinetics with the rate constant of elimination k_e.

It clearly appears that the development of drugs, and especially of the dosage forms through which they are delivered, is facilitated and enhanced after pharmacokinetic measurements have been taken.

1.4 PHARMACODYNAMICS

Pharmacodynamics is concerned with what the drug does to the body, dealing with the question of what happens to the patient's organism under the influence of the drug. Thus, its objective is the therapeutic effect on the patient.

1.5 LIBERATION OF THE DRUG

The term *liberation* describes the release of the drug along the GIT from the dosage form that has been administered. The site of application, which is either the stomach (the gastric liquid is acid with a pH around 2) or along the intestinal tract (slightly basic), can be selected by using special coatings that are unchanged in the gastric liquid. Thus, the factors determining the process of release of the drug largely depend on the nature of the excipient or of the protective coating.

In the case of immediate release obtained with a fast rate of dissolution or disintegration of the excipient, these factors are the solubility and the rate of dissolution of the drug. At this point, the terms *hydrophilicity* and *lipophilicity* should be briefly considered. Solubility can be obtained between either the hydrophilic or the lipophilic phase. Hydrophilicity characterizes the tendency to aqueous or polar molecules to bind the drug molecule; lipophilicity represents the affinity of a drug molecule for a lipophilic environment. A distribution coefficient between these two phases is a critical parameter. The rate of dissolution is the second important factor, but except for particular food problems, it is faster than the rate of absorption in the blood.

In the case of controlled (sustained) drug release dosage forms, the process of release is quite different. The rate of drug release, controlled by either a diffusion or erosion process, is very low and intended to be far lower than the rate of absorption of the drug in the blood.

1.6 MEMBRANES

After its liberation from the dosage form along the GIT, the drug must pass through various barriers before reaching the site of therapeutic action. All these barriers consist of membranes, so it is useful to describe them succinctly.

Membranes play the role of biological interfaces or barriers between morpho-logical and functional entities. Following the model that still holds, they consist of bimolecular layers of 5.5 nm long (55 Å) phospholipids, whose polar ends are covered by a 3-nm-thick layer of protein molecules on each face, leading to a membrane of a thickness around 11 nm [17]. The bimolecular lipids, lying as open parallel fibers covered on both sides by protein, point their hydrophilic ends toward the extracellular and intracellular spaces.

Membranes are also involved in many specific transports of drug mechanisms. The process of transport is controlled by diffusion, resulting from a transmembrane movement of the drug following the concentration gradient across the membrane thickness. Thus the dissolved drug passes into the membrane at a rate that depends on the thickness of the membrane, the dimension of the drug, and its solubility into the membrane.

1.7 BIOAVAILABILITY

The term *bioavailability* refers to the amount of the drug absorbed by the organism that arrives at the target site in a given time. It is defined as the percentage of the drug absorbed from a given dosage form.

This concept was introduced by Oser et al. [18] as "physiological availability." More recently [19], it was defined as the ratio of the amount of unchanged drug reaching the circulating system after administration of a test dose to that after administration of a standard dosage form. Bioavailability has also been determined using the drug level in the blood or even using urine excretion profiles.

Several factors intervene in lowering the proportion of the drug able to reach the systemic circulation: incomplete absorption through the membrane of the GIT; the entire dose must pass through the liver, and this first-pass hepatic is responsible for a liver extraction and a presystemic metabolism, which is a major factor in reducing the bioavailability of lipophilic drugs; an additional first-pass effect leading to metabolism may occur in the GIT itself.

Absolute bioavailability can be calculated by comparing the results obtained by administration of a drug orally and intravenously, and measuring the drug concen-trations at various times in both cases. The ratio of the *area under the curve* (AUC) obtained gives a value of this absolute bioavailability. The area under the curve is calculated by integrating with respect to time the profile of the drug concentration previously measured at various times:

$$AUC_{0-t} = \int_0^t C_t \, dt \qquad (1.1)$$

The bioavailability is thus given by the ratio:

$$\frac{AUC_{oral}}{AUC_{i.v.}} \qquad (1.2)$$

1.8 ABSORPTION OF THE DRUG IN THE BLOOD

Absorption is the assimilation of the drug from the GIT into the bloodstream or the lymphatic system. Molecules of drug must pass through the various complex membranes made of lipid barriers, and various mechanisms may be involved in the process of absorption. The following stages must be considered:

- Dissolution of the drug into the membrane material
- Transcellular passive diffusion through the membrane, or even active transport
- Luminal and epithelial metabolism

In terms of mathematics, the transport of the drug from the GIT into the systemic system is generally expressed by a first-order kinetics. Thus the amount of drug in the GIT decreases according to the following equation:

$$M_t = M_{in} \exp(-k_a t) \tag{1.3}$$

This leads to a simultaneous increase in the concentration of the drug in the blood:

$$C_t = C_\infty [1 - \exp(-k_a t)] \tag{1.4}$$

where the concentration of free drug not bound to the plasmatic protein is given by:

$$C_\infty = \frac{M_{in}}{V_p} \tag{1.5}$$

The kinetics of absorption of various drugs is shown in Figure 1.2, by evaluating the decrease in the amount of drug remaining along the GIT and the corresponding increase in the drug concentration in the blood. The three values of the rate constant of absorption are those found for various drugs such as ciprofloxacin, cimetidine, and Aspegic[R] [20]. The effect of the value of the rate constant of absorption associated with the nature of the drug for a given patient is demonstrated in Figure 1.2 when the rate constant is largely varied, from 1.3/h for ciprofloxacin to 2.9/h for Aspegic.

The rate constant of absorption k_a can be calculated from the drug profile obtained in the blood as a function of time by considering either the whole curve or the characteristics of the maximum concentration in this curve (time, height), as is described in detail with oral dosage forms with immediate release (Chapter 3). The time at which the maximum of the drug concentration passes through is given in the Equation 1.6, where the two rate constants of absorption and elimination k_a and k_e intervene:

$$t_{max} = \frac{1}{k_a - k_e} Ln \frac{k_a}{k_e} \tag{1.6}$$

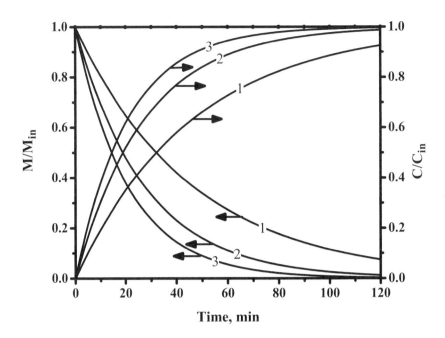

FIGURE 1.2 Effect of the drug's rate constant of absorption on the kinetics of absorption of that drug in the blood (the time is expressed in minutes). Left: amount of drug remaining along the GIT vs. time. Right: corresponding increase in the concentration of the drug in the blood. 1. Ciprofloxacin with k_a = 1.3/h. 2. Cimetidine with k_a = 2.2/h. 3. Aspegic with k_a = 2.9/h.

1.9 DISTRIBUTION OF THE DRUG

The organism consists of cells and fluid. The body fluid can be divided into three compartments:

- Intravascular compartment
- Interstitial or extracellular compartment
- Intracellular compartment

Distribution is a term indicating the way in which the unbound drug in the blood passes into the tissues or organs. The drug molecule leaves the intravascular compartment to be redistributed between extracellular and intracellular compartments, and reaches the receptors lying in the tissues.

1.10 ELIMINATION OF THE DRUG

The drug is inactivated by various processes, based on chemical alteration of the molecule with metabolism, and excretion via various organs.

1.10.1 METABOLISM

Metabolism is the sum of all chemical reactions involved in the biotransformation of endogenous and exogenous substances that occur in the living cells. There are many possible processes of biotransformation. Catabolic reaction breaks the drug into simpler substances. Anabolism is the synthesis of new molecules from simpler substances. Most stages require the presence of enzymes as catalysts.

The metabolites formed from a drug during its biotransformation are generally more water-soluble than the drug itself, facilitating their excretion through the urine.

1.10.2 EXCRETION AND CLEARANCE VIA VARIOUS ORGANS

Excretion of the drug and its metabolites from the organism is primarily undertaken through the kidney and the intestinal tract; only gaseous substances (anesthetics) are excreted via the lungs. The major route of excretion is through the kidneys, and the rate and extent of renal excretion is determined by glomerular filtration, tubular reabsorption, and tubular secretion.

A useful pharmacokinetic parameter is *clearance* (*Cl*). Clearance of the drug occurs by the perfusion of the blood to the organs of extraction. Extraction (*E*) refers to the proportion of the drug that is excreted or altered (metabolism); it is related to the flow of blood *Q* through the organ of excretion by the formula:

$$Cl = QE \tag{1.7}$$

Because the organs of extraction are mainly the liver, with the hepatic clearance Cl_h, and the kidney, with the renal excretion (renal clearance Cl_r), the overall value of the systemic clearance Cl is their sum:

$$Cl = Cl_h + Cl_r \tag{1.8}$$

In other words, clearance is expressed in terms of the blood volume circulating through the organ of excretion necessary to get free of the drug per unit time:

$$\text{Rate of excretion} = Cl\ C \tag{1.9}$$

where the rate of excretion is the amount of drug eliminated per unit time.

1.10.3 BIOLOGICAL HALF-LIFE TIME OF THE DRUG

Half-life of elimination of the drug is a main pharmacokinetic parameter. It represents the time necessary for the concentration of the unbound drug in the blood to be reduced by half from the initial value. As the rate of elimination can be described by a first-order kinetics with the rate constant of elimination k_e, the general equation for the elimination of the drug

$$\frac{dC}{dt} = -k_e C \tag{1.10}$$

gives after integration

$$Ln\frac{C_t}{C_0} = -k_e t \tag{1.11}$$

which leads to the expression for the time of half-life:

$$Ln2 = k_e t_{0.5} = 0.693 \tag{1.12}$$

It is worth noting that Equation 1.11 can be rewritten in the same form as shown in Equation 1.3:

$$C_t = C_0 \exp(-k_e t) \tag{1.13}$$

Two other relations are well known for expressing clearance. They show that the amount of drug in circulation is related to the volume of distribution and to the rate constant of elimination. The one relating the amount of drug present in the blood after administration, or the concentration of the drug in the apparent volume of distribution, in terms of the *AUC*, leads to:

$$Cl = \frac{Dose}{AUC} = \frac{CV_p}{AUC} \tag{1.14}$$

The other expresses that the amount of drug in circulation is related to the volume of distribution and to the rate constant of elimination:

$$Cl = k_e V_p \tag{1.15}$$

This shows that the clearance is the apparent volume of distribution getting free of drug per unit time.

The effect of the nature of the drug, with its rate constant of elimination, clearly appears in Figure 1.3, where the rate constant is largely varied, from 0.004/h for tamoxifene to 0.34/h for cimetidine. This fact will be considered in detail in the following chapters.

1.11 THERAPEUTIC INDEX

Ehrlich introduced the concept of the *therapeutic index,* defined as the relationship between the minimum curative dose and the maximum tolerated dose. In pharmacology, the therapeutic index is the ratio relating the median lethal dose (the concentration causing the deaths of 50% of experimental animals) and the median effective dose (the concentration at which the drug is effective in 50% of cases):

$$TI = \frac{LD_{50}}{ED_{50}} \tag{1.16}$$

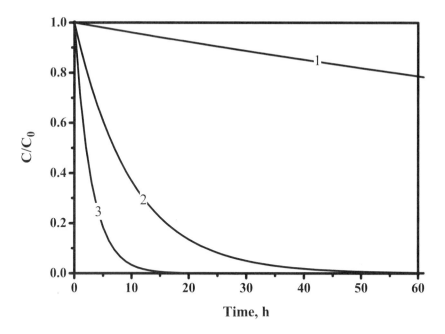

FIGURE 1.3 Effect of the rate constant of elimination on the kinetics of elimination of the drug from the body (time is expressed in hours). 1. Tamoxifene with $k_e = 0.004$/h. 2. Ciprofloxacin with $k_e = 0.12$/h. 3. Cimetidine with $k_e = 0.34$/h.

Of course, the higher the therapeutic index is, the greater the safety margin.

This concept of therapeutic index comes from the *threshold model*, based on the assumption that there is an exposure concentration, or threshold, for the drug that is considered to be active. It is worth noting that a linear model also exists, meaning that the response to the dose becomes zero when the dose equals zero. Also, the theory of hormesis suggests that there is some stimulatory and possibly advantageous effect from low doses of a substance that is toxic at higher levels [21].

1.12 DETERMINATION OF THE PHARMACEUTICAL PARAMETERS

At this point, it is necessary to trace how to make measurements and calculations in order to obtain the values of the relevant pharmacokinetic parameters:

- The rate constant of elimination k_e
- The apparent volume of distribution (which is often much larger than the volume of the blood or plasma) V_p
- The systemic clearance Cl
- The rate constant of absorption k_a

1.12.1 RATE CONSTANT OF ELIMINATION

After an intravenous injection of a given dose, the concentration of the drug in the blood C_0 cannot be measured at time 0 for at least two reasons. One reason is that it takes some time for the drug to be distributed throughout the body. This first phase, short in time, is called the *phase of distribution*, with a rate constant and the corresponding half-life time $t_{0.5\alpha}$. The other reason is that it is not possible to measure the concentration of the drug at the time of injection. This phase of distribution, very short in time, appears only with intravenous injection in bolus and not with oral drug administration. There is a small difference in the AUC in finding by extrapolation the initial concentration C_0 by considering only the phase of elimination (with the half-life time $t_{0.5\beta}$, denoted $t_{0.5}$). However, the drug concentration can easily be measured at any time. Thus, two values of this concentration measured at two different times enable one to apply Equation 1.13:

$$C_1 = C_0 \exp(-k_e t_1)$$
$$C_2 = C_0 \exp(-k_e t_2)$$

(1.17)

The unknown concentration C_0 is eliminated by dividing the preceding relations by themselves:

$$\frac{C_1}{C_2} = \exp[-k_e(t_1 - t_2)]$$

(1.18)

and k_e is obtained:

$$k_e = \frac{1}{t_2 - t_1} Ln \frac{C_1}{C_2}$$

(1.19)

The initial concentration C_0 at time 0 is thus given by

$$C_0 = C_1 \exp(k_e t_1) = C_2 \exp(k_e t_2)$$

(1.20)

For instance, by making the ratio of the two concentrations C_1 and C_2 measured at time t_1 and t_2, respectively, the value for k_e is obtained by using Equation 1.19:

$$k_e = 0.1/h$$

whereas the initial concentration C_0 can be evaluated using Equation 1.20.

1.12.2 VOLUME OF DISTRIBUTION (APPARENT VOLUME V_p)

The apparent volume of distribution is thus obtained from measurements made with intravenous injection. After calculating the initial concentration of the unbound drug in the blood at time 0, C_0, V_p is obtained by using Equation 1.21, where *Dose* represents the amount of the drug injected:

$$V_p = \frac{Dose}{C_0} \qquad (1.21)$$

Notice the similarity of Equation 1.5 and Equation 1.21, written in two different cases: the former in the case of absorption (Section 1.8) and the latter in the present case of i.v. injection.

1.12.3 RATE CONSTANT OF ABSORPTION

As shown in Section 1.8, the rate constant of absorption k_a is obtained by using Equation 1.6 and iterative calculation in order to fit the terms of Equation 1.22 that derived from Equation 1.6:

$$(k_a - k_e)t_{max} = Ln\frac{k_a}{k_e} \qquad (1.22)$$

1.12.4 SYSTEMIC CLEARANCE

The systemic clearance Cl, which is the sum of the renal and hepatic clearances, can be obtained from measurements made with i.v. injection by using Equation 1.14, or more simply with Equation 1.15. The rate constant of elimination k_e characterizes the whole process of elimination of the drug from the blood, including renal and hepatic contributions, provided that these two processes obey a first-order kinetics.

1.12.5 LINEAR OR NONLINEAR PHARMACEUTICS

The pharmaceutics for a drug is said to be *linear* when the concentration of the drug in the blood is proportional to the dose, whatever the method of administration, either with oral dosage forms or through intravenous injection or infusion. This means that the processes of absorption, metabolism, and elimination—as well as the process in which the drug binds the plasmatic proteins—are controlled by first-order kinetics, and that the rate constant of these processes remains constant whatever the dose or, more often, within a limited range of the dose. In the same way, the plasmatic volume is constant.

In other cases, the pharmacokinetics is said to be *nonlinear*. For instance, the rate constant of elimination for ethanol varies with the dose. Various reasons cause nonlinearity: a change in the kinetics and the rate constant of absorption, a change in the processes of distribution or of biotransformation, or a change in the kinetics and the rate of elimination.

Concerning the use of oral dosage forms with sustained release, the pharmacokinetics of the drug should be linear. Thus, the drug is changed in the patient's body in the same way however it is delivered, with immediate release or sustained release.

REFERENCES

1. Siepmann, J. et al., Calculation of the dimensions of polymer-drug devices based on diffusion parameters, *J. Pharm. Sci.,* 87, 827, 1988.
2. Fessi, H. et al., Square-root of time dependence of matrix formulations with low drug content, *J. Pharm. Sci.,* 71, 749, 1982.
3. Brossard, C. et al., Dissolution of a soluble drug substance from vinyl polymer matrices, *J. Pharm. Sci.,* 72, 162, 1983.
4. Wouessidjewe, D. et al., Préparation de microgranules de trinitrine à libération prolongée par pelliculage en milieu aqueux, *Labo. Pharma. Probl. Tech.,* 32, 772, 1984.
5. Droin, A. et al., Model of matter transfers between sodium salicylate–Eudragit matrix and gastric liquid, *Int. J. Pharm.,* 27, 233, 1985.
6. Batyrbekov, E.O. et al., Some fields of biomedical application of polyurethanes, *Brit. Polym. J.,* 23, 273, 1990.
7. Bawa, R. et al., An explanation for the controlled release of macromolecules from polymers, *J. Control. Release,* 1, 259, 1985.
8. Yean, L., Bunel, C., and Vairon, J.P., Reversible immobilization of drugs on a hydrogel matrix. Diffusion of free chloramphenicol from poly(2-hydroethyl methacrylate) hydrogels. *Makrom. Chem.,* 191, 1119, 1990.
9. Touitou, E. and Donbrow, M., Drug release from non-disintegrating hydrophilic matrices with Na salicylate as model drug, *Int. J. Pharm.,* 11, 355, 1982.
10. Peppas, N.A. et al., Modelling of drug diffusion through swellable polymeric systems, *J. Membrane Sci.,* 7, 241, 1980.
11. Malley, I. et al., Modelling of controlled release in case of Carbopol–sodium salicylate matrix in gastric liquid, *Int. J. Pharm.,* 13, 67, 1987.
12. Peppas, N.A. and Bindschaedler, C., Les dispositifs à libération contrôlée pour la délivrance des principes actifs médicamenteux. IV. Systèmes à gonflement contrôlé, *STP. Pharma.,* 2, 38, 1986.
13. Heller, J., Biodegradable polymers in controlled drug delivery. *CRC Critical Reviews in Therap. Drug Carrier Systems,* 1, 39, 1984.
14. Davis, S.S., The design and evaluation of controlled release systems for the gastrointestinal tract, *J. Control. Release,* 2, 27, 1985.
15. Stockwell, A.F., Davis, S.S., and Walker, S.E., In vitro evaluation of alginate gel systems as sustained release drug delivery systems, *J. Control. Release,* 3, 167, 1986.
16. Ferrari, F. et al., In vivo evaluation of gastro adhesive granules intended for site-specific treatment of *Helicobacter pylori* related pathologies, in *Proc. 3rd World Meeting on Pharmaceutics, Biopharmaceutics and Pharmaceutical Technology,* APV/APGI, Berlin, 2000, 873.
17. Davson, H. and Danielli, J.F., *The Permeability of Natural Membranes,* 2nd ed. 1952, Cambridge University Press, New York.
18. Oser, B.L., Melnick, D., and Hochberg, M., *Ind. Eng. Chem. Anal. Chem. Educ.,*17, 405, 1945.
19. Levy, G., Gibaldi, M., and Jusko, J., Multicompartment pharmacokinetic models and pharmacologic effects, *J. Pharm. Sci.,* 58, 422, 1969.
20. Vergnaud, J.M., Use of polymers in pharmacy for oral dosage forms with controlled release, *Recent. Res. Devel. Macromol. Res.,* 4, 173, 1999.
21. Hogue, C., Low-dose effects, *Chemical Eng. News,* April 5, 2004.

2 Intravenous Administration

NOMENCLATURE

AUC_{0-T}, AUC_{T-2T} Area under the curve up to time T, between T and $2T$, respectively.

C_0 Initial free drug concentration in the blood corresponding with the dose injected.

C_t, $C_{0.5}$ Free drug concentration in the blood at time t, at time $t_{0.5}$, respectively.

C_n^{max}, C_n^{min} Drug concentration at the nth peak, at the nth trough, respectively.

C_∞ Free drug concentration in the blood, under steady state with infusion.

C_{ti} Plasma drug concentration at time t_i.

Dose Amount of drug delivered intravenously in the blood.

k_e Rate constant of elimination of the drug, expressed per hour (/h).

T Interval of time between two injections.

t, $t_{0.5}$, t_i Time, half-life time of the drug, time of drug infusion.

$\theta = \frac{t}{t_{0.5}}$ Dimensionless number for the time.

U Rate of administration of the drug with infusion (mass per unit time).

V_d **or** V_p Apparent volume of distribution or volume of blood (l/kg).

Intravenous administration (i.v.) is an effective way to deliver the drug in the patient's body. Although it is not the purpose of this book to describe these methods deeply, they are presented briefly in the present chapter, by considering the administration of a single dose and of multidoses in succession, and finally the continuous administration with a constant rate of drug (infusion).

2.1 ADMINISTRATION OF A SINGLE DOSE

When the drug is administered intravenously into the circulation, the compound undergoes distribution into tissues before clearance, as shown in Figure 2.1. For a drug undergoing rapid distribution, the following three pharmacokinetic parameters, which are necessary to describe the process precisely, can be obtained from a single compartment model:

- Half-life time of the drug in the body
- Apparent volume of distribution
- Systemic clearance

They are evaluated from the experimental curve describing the decrease in the concentration of the free drug in the blood with time, by making the mathematical treatment that follows.

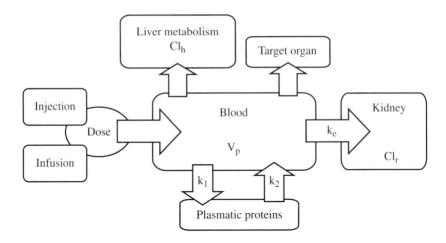

FIGURE 2.1 Path followed by the drug with intravenous delivery.

For the dose injected at time 0, responsible for the initial concentration C_0 of the free drug present in the circulation, the ratio of these is the volume of distribution V_p:

$$V_p = \frac{Dose}{C_0} \tag{2.1}$$

Because of the distribution of the drug to various tissues and organs, as well as the binding to plasmatic proteins, this volume can greatly exceed the blood volume. The pharmacokinetic parameters associated with the decrease in the concentration of the free drug in the blood are evaluated by using Equation 2.2, where k_e is the rate constant of elimination and C_0 is the value of the initial concentration of this drug at the time of injection, shown in Equation 2.1:

$$C_t = C_0 \exp(-k_e t) \tag{2.2}$$

The rate constant of elimination of the drug and the unknown initial concentration are calculated from the values of the concentration measured at times t_1 and t_2, as indicated in Chapter 1 (1.12.1), by using the following two equations:

$$k_e = \frac{1}{t_2 - t_1} Ln \frac{C_1}{C_2} \tag{2.3}$$

$$C_0 = C_1 \exp(k_e t_1) = C_2 \exp(k_e t_2) \tag{2.4}$$

The half-life time of the drug in the blood is easily obtained by using the Equation 2.2, in putting the concentration C_t equal to the half of the initial

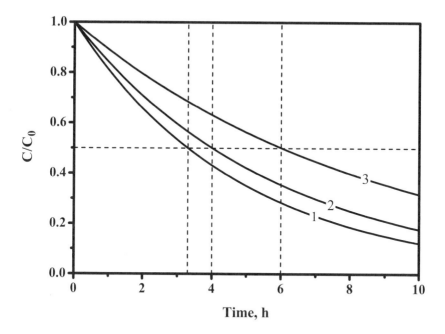

FIGURE 2.2 Kinetics of elimination of the drug with intravenous injection of ciprofloxacin with various values of k_e. 1. 0.21/h. 2. 0.173/h. 3. 0.115/h.

concentration C_0:

$$Ln\frac{C_{0.5}}{C_0} = Ln(0.5) = 0.693 = k_e t_{0.5} \qquad (2.5)$$

Figure 2.2 shows the decrease in concentration of the free drug in the blood as a function of time, calculated for three values of the rate constant of elimination. Some of the extreme values of this rate constant of elimination obtained for ciprofloxacin with various patients are collected in Table 2.1.

The strong differences among these profiles of drug concentration make clear that the selection of the pharmacokinetic parameters required to determine the dose for unknown patients is of great concern, making necessary a precise determination

TABLE 2.1
Pharmacokinetic Parameters of Ciprofloxacin

V_p (l/kg)	2–3	1.7–2.7			
$t_{0.5}$ (h)	4–6	2.9–3.7	3.1–4.9	2.2–7.8	2.7–4.5
k_e (/h)	0.17–0.12	0.24–0.19	0.22–0.14	0.31–0.09	0.26–0.15
References	[1]	[2]	[3,4]	[5]	[6]

of the value of the half-life time for each patient before starting the therapy. Furthermore, it clearly appears that healthy and unhealthy patients show different behaviors toward the drug, as is especially shown for ciprofloxacin.

2.2 REPEATED INTRAVENOUS INJECTION (DIV)

When bolus i.v. doses are administered far apart in time, they behave independently, as shown in Figure 2.2, when they are alone. But this is not the most desirable profile of concentration, because a certain minimum concentration for the troughs is needed to maintain efficacy, and the peaks should not exceed a maximum value to prevent the occurrence of side effects. Deliveries via bolus i.v. should thus be sufficiently close together so that subsequent doses are administered prior to the full elimination of the preceding doses and some accumulation will develop. As a result, a so-called *steady state*, meaning that the profiles are exactly reproduced by each subsequent dose, is attained after a few injections. This fact is observed in Figure 2.3, where the drug concentration at time t, C_t as a fraction of the initial concentration C_0, is drawn as a function of time, for repeated doses with the same interval of time equal to the half-life time of each drug, with various values of this half-life time: 3.3 h in curve 1 and 6 h in curve 2.

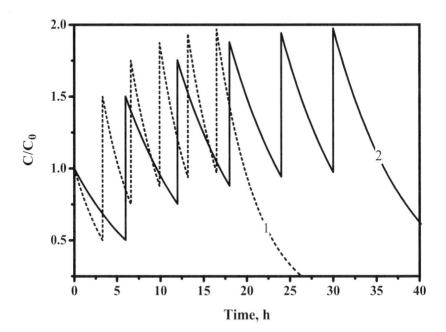

FIGURE 2.3 Profile of the free drug concentration in the blood with repeated doses injected at the same interval of time, with two values of the rate constant of elimination of ciprofloxacin as the drug and the interval of time equal to the half-life time. 1. $k_e = 0.21/h$ and $t_{0.5} = 3.3$ h. 2. $k_e = 0.115/h$ and $t_{0.5} = 6$ h.

2.2.1 CALCULATION OF THE DRUG CONCENTRATION AT PEAKS AND TROUGHS

Calculation is made by taking care that each dose provokes a drug concentration in the blood that comes in addition to the remaining concentration from previous doses. When the pharmacokinetics is linear, this law of addition for the doses and their associated concentrations is right, meaning that the concentration of the drug in the body is proportional to the dose injected. By introducing the interval of time T between each subsequent dose, and by putting $\theta = \frac{T}{t_{0.5}}$, Equation 2.2 is written in the following form:

$$C_T = C_0 \exp(-k_e T) = C_0 \exp(-0.693\theta) \tag{2.6}$$

The principle of calculation is as follows. At the time θ (dimensionless time) associated with the first dose, the concentration is given by Equation 2.6. The second dose injected at that time T causes the new drug concentration to reach the first peak

$$C_T = C_0 + C_0 \exp(-0.693\theta) = C_0(1 + \exp(-0.693\theta)) \tag{2.7}$$

The second peak is reached when the third dose is administered at time $2T$ (or $2\,\theta$):

$$C_{2T} = C_0 + C_0(1 + \exp(-0.693\theta))\exp(-0.693\theta) \tag{2.8}$$

which can be rewritten in terms of a series:

$$\frac{C_{2T}}{C_0} = 1 + \exp(-0.693\theta) + \exp^2(-0.693\theta) \tag{2.9}$$

For a number of doses n, the series becomes

$$\frac{C_{nT}}{C_0} = \sum_0^n \exp^n(-0.693\theta) = \sum_0^n \exp^n(Ln(0.5)^\theta) = \sum_0^n (0.5)^{\theta n} \tag{2.10}$$

This is a convergent series whose terms follow a geometrical progression from 1 to 0, and the sum of this series is well known, so that Equation 2.10 is finally resolved when n is so large that it can be considered infinite. Thus, after a sufficient number of repeated doses for the steady state to be attained, the concentration is given by

$$\frac{C_{nT}}{C_0} = (1 - (0.5)^\theta)^{-1} \tag{2.11}$$

TABLE 2.2
Values of the Peaks Concentration C_n^{max} as a Function of n and θ

θ /n	1	2	3	4	5	6	7	8	9	10	∞
1	1.5	1.75	1.87	1.94	1.97	1.98	1.992	1.996	1.997	1.998	2
1.5	1.35	1.48	1.52	1.54	1.545						1.547
2	1.25	1.31	1.328	1.332	1.333						1.333
3	1.125	1.14	1.142	1.143							1.143

If we remark that the concentrations shown in Equation 2.10 and Equation 2.11 are the concentrations of the peaks, it is easy to get the extreme values of the peaks and troughs through which the drug concentration alternates. The concentration of the peak increases with n according to

$$\frac{C_n^{max}}{C_0} = \sum_0^n (0.5)^{\theta n} \tag{2.12}$$

The concentration of the following troughs is given by the obvious relation:

$$\frac{C_n^{min}}{C_0} = -1 + \sum_0^n (0.5)^{\theta n} \tag{2.13}$$

The values of the maximum values of the drug concentration in the body as a function of the number of DIV are shown in Table 2.2, for four values of the interval of time T between doses expressed in terms of the half-life time θ and for various values of repeated injections n.

The values of the minimum values of the drug concentration in the body as a function of the number of DIV for the same values of the ratio of time θ are easily obtained by using the obvious relationship:

$$C_n^{min} = C_n^{max} - C_0 \tag{2.14}$$

2.2.2 Effect of the Nature of the Drug

The effect of the nature of the drug and of the value of its half-life time clearly appears in the Figure 2.2 and Figure 2.3. The three curves drawn in Figure 2.2 are calculated with some extreme values of the rate constant of elimination of ciprofloxacin, ranging from 0.115/h to 0.21/h, and the half-life time varied from 3.3 h to 6 h. The two curves in Figure 2.3 represent the profiles of concentration obtained with ciprofloxacin when the dose is administered in succession at the same interval of time T, taken as $\frac{1}{2} t_{0.5}$.

The following two results are worth noting. The one, more evident, is that the time of duration of the *DIV* is shorter when the value of the rate constant is larger and the half-life time shorter, with the obvious statement: the larger the rate constant of elimination, the shorter in time the delivery of the drug. The other result, more important perhaps, is that the concentration of the drug alternates between higher peaks and higher troughs when the rate constant of elimination is larger. Thus, when the doses are delivered at the interval of time T taken as equal to the half-life time, the other statement holds: the larger the rate constant of elimination (or the shorter the half-life time), the higher the peaks and troughs.

Because a drug is associated with its typical pharmacokinetic parameters, normally it would be necessary to calculate and draw for each drug the profile of concentration as a function of time in the case of repeated injections. This inconvenience can be avoided by using a master curve with dimensionless coordinates. Two master curves are shown in Figure 2.4, where the concentration of the drug as a fraction of the initial concentration C_0 is expressed in terms of the dimensionless time θ equal to $\frac{t}{t_{0.5}}$. Two curves are drawn in Figure 2.4: the dotted curve (1) is obtained when the interval of time T between each injection is equal to the half-life time of the drug, whereas curve 2 is drawn when this interval of time T is 1.5, the half-life time of the same drug. The effect of the interval of time on the profile of the drug concentration clearly appears in this figure, which can be defined by the following statement: the shorter the interval of time between every dose, the higher the concentrations of the drug, with a special mention of peaks and troughs.

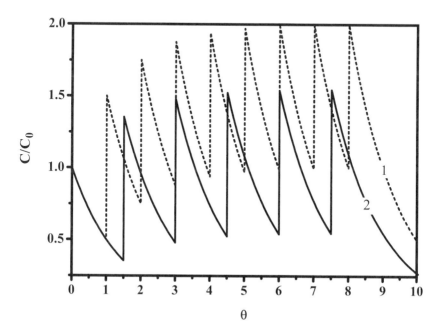

FIGURE 2.4 Master curve for the profile of the drug concentration in the blood with repeated doses at various intervals of time, with ciprofloxacin as the drug. 1. Interval of time $= t_{0.5}$ of the drug. 2. Interval of time $= 1.5\ t_{0.5}$ of the drug.

The protocol for using the master curves shown in Figure 2.4 is as follows. When the half-life time is known (e.g., 6 h), θ representing the dimensionless time $\frac{t}{t_{0.5}}$ can be associated with normal time (in hours); for $\theta = 10$ it corresponds to 60 hours. Thus, each dose is injected at the interval of 6 h in case of the curve 1 ($\theta = 1$) and of 9 h in case of the curve 2 ($\theta = 1.5$). The concentration of the drug in these two cases, at any time and especially for the peaks and troughs, is expressed in terms of the concentration C_0 defined by Equation 2.1. The precise values at the peaks and troughs are given in Table 2.2 for the subsequent doses delivered in succession (with the number n) and for various values of the interval of time T, expressed in terms of the half-life time through the dimensionless time θ.

2.2.3 Calculation of the Area under the Curve

The area under the curve (AUC) is evaluated by using the relations that follow.
For a single injection, it is easily obtained:

$$AUC_{0-\infty} = C_0 \int_0^\infty \exp(-k_e t)dt = C_0 \frac{t_{0.5}}{0.693} \qquad (2.15)$$

For two doses injected in succession at the interval of time T, the AUC is the sum of the integrals:

$$AUC_{0-\infty} = AUC_{0-T} + AUC_{T-\infty} = C_0 \int_0^T \exp(-k_e t)dt + C_1^{\max} \int_T^\infty \exp(-k_e t)dt \qquad (2.16)$$

For n injections, the recurrent relation is obtained with the corresponding integrals:

$$AUC_{0-\infty} = AUC_{0-T} + AUC_{T-2T} + \quad AUC_{(n-1)T-nT} + AUC_{nT-\infty} \qquad (2.17)$$

It is of interest to predict by calculation the value of the AUC associated with the system of n injections, when the half-life time of the drug is known, for various values of the interval of time T of the successive injections. In Figure 2.3 or Figure 2.4, where the profile of the drug concentration is drawn, the successive areas corresponding to the amounts of drug delivered in the body over each period of time T can be evaluated as follows.

For the first period of time, when the initial concentration at time 0 is C_0 in the body, the amount is obtained by integrating the concentration $C_t dt$ from 0 to T:

$$AUC_{0-T} = C_0 \int_0^T \exp(-k_e t)dt = \frac{C_0}{k_e}[1 - \exp(-0.693\theta)] = \frac{C_0}{k_e}[1 - (0.5)^\theta] \qquad (2.18)$$

In the same way, within the interval of time between T and $2T$, the AUC is

$$AUC_{T-2T} = C_1^{max} \int_0^T \exp(-k_e t)dt = \frac{C_1^{max}}{k_e}[1-\exp(-1.693\theta)] = \frac{C_1^{max}}{k_e}[1-(0.5)^\theta]$$

(2.19)

where C_1^{max} is the peak concentration obtained at time T after the second injection.

The following recurrent relationship allows calculation of the amount of drug in the body over any period of time, for example, $(n-1)T$ and T, using the corresponding maximum value of the initial drug concentration in the body C_{n-1}^M:

$$AUC_{(n-1)T-nT} = C_{n-1}^{max} \int_0^T \exp(-k_e t)dt = \frac{C_{n-1}^{max}}{k_e}[1-\exp(-0.693\theta)] = \frac{C_{n-1}^{max}}{k_e}[1-(0.5)^\theta]$$

(2.20)

Finally, the amount of drug that passed into the body during n successive injections delivered with a period of time T, is given by Equation 2.17, which becomes

$$AUC_{0-\infty} = \frac{C_n^{max}}{k_e} + \frac{1}{k_e}[1-(0.5)^\theta]\sum_0^{n-1} C_n^{max}$$

(2.21)

The partial AUC associated with every successive dose, given by Equation 2.20, is expressed as a fraction of the total AUC obtained with a single dose taken alone, shown in Table 2.3 for the two values $\theta = 1$ and $\theta = 1.5$.

The AUC obtained with a single dose is given in Equation 2.15. This calculation is made by using the data shown in Table 2.2 for the maximum values at peaks for $\theta = 1$ and 1.5, respectively. A similar calculation can be made for other values of the interval of time T by calculating the value of the concentration at the following peaks by using Equation 2.12 for the desired value of θ. The total AUC is the sum of each partial AUC, and when n is large, it is nth times that obtained with the single dose, with a slight change in the last one.

TABLE 2.3
Values of AUC for Each Multidose as a Fraction of the AUC with a Single Dose

Dose Number	1	2	3	4	nth	Last
$\theta = 1$	0.5	0.75	0.875	0.94	1	2
$\theta = 1.5$	0.65	0.87	0.96	0.98	1	1.547

2.2.4 CHANGE IN THE DOSING DURING THE DIV

Of course, i.v. therapy exhibits some drawbacks in comparison with oral dosage forms, the essential one being that it is generally administered at the hospital. Nevertheless, an advantage exists for i.v. administration, because a change in the conditions of the treatment is possible, by varying either the dose injected or the interval of time between subsequent injections. Depending on the response of each patient to the dose of the drug, as shown for ciprofloxacin in Table 2.1, the values of the two main parameters V_p and $t_{0.5}$ (apparent volume of distribution and half-life time) range within a wide extent. Thus, it seems necessary to deal with this problem in adapting the right operational conditions of the therapy during treatment.

The protocol may be as follows. Consider curves 1 and 2 in Figure 2.4. After the first injection with the predetermined dose (associated with the hypothetic initial concentration C_0), whatever the interval of time T and the associated dimensionless time θ, two analyses are made at two times chosen between 0 and T. These two measurements enable the therapist to calculate the actual initial concentration C_0 and the half-life time of the drug for the patient by using the Equation 2.1, Equation 2.3, and Equation 2.4. Then, two cases may occur: one with the apparent volume of distribution and the other with the half-life time.

When the value obtained for the volume of distribution is different from the one that was estimated in selecting the first dose, the second and subsequent doses may thus be adjusted by increasing or decreasing their value, using Equation 2.1.

When the value of the half-life time obtained by calculation is quite different from the one that was initially expected, the curves in Figure 2.4 help the therapist in selecting the new interval of time T between successive doses. For instance, when the desired therapy should be done with $T = t_{0.5}$, the interval of time is adjusted to the value of the half-life time obtained with this calculation. The effect of a change in the dose and the corresponding concentration of the drug in the blood can be appreciated in Figure 2.5, where the second and subsequent doses are either 20% more or less than the first dose, when the interval of time between two successive doses is equal to the half-life time of the drug.

2.3 CONTINUOUS INTRAVENOUS INFUSION (CIV)

2.3.1 GENERAL EXPRESSION OF CIV

The general equation for continuous intravenous (CIV) infusion, taking into account the rate of drug delivered in the body through infusion and the rate of drug eliminated, is written as follows:

$$\frac{dM}{dt} = U - k_e M \tag{2.22}$$

where M is the mass of drug and U the mass of drug administered per unit time.

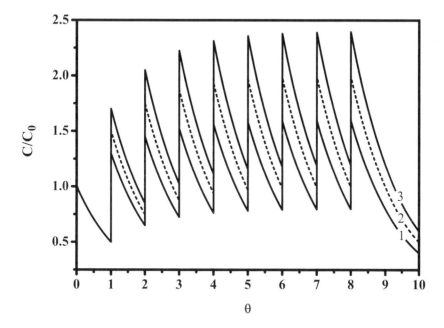

FIGURE 2.5 Master curve for the concentration profile of the free drug in the blood with repeated doses injected at constant interval of time equal to the half-life time when the subsequent doses differ by 20% from the initial dose responsible for the initial concentration C_0 at time 0. 1. The second and subsequent doses are 20% lower than the first dose. 2. Constant doses associated with the constant concentrations equal to C_0. 3. The second and subsequent doses are 20% larger than the first dose.

By dividing both terms by the volume of distribution V_p, the preceding equation is expressed in terms of concentration of the drug:

$$\frac{dC}{dt} = \frac{U}{V_p} - k_e C \tag{2.23}$$

After integration between 0 and t, the fundamental expression for the concentration in the blood as a function of time is obtained, related to the pharmacokinetic parameters (e.g., the rate constant of elimination and the apparent volume of distribution):

$$C_t = \frac{U}{k_e V_p}[1 - \exp(-k_e t)] \tag{2.24}$$

As shown in Equation 2.24, the concentration of the free drug in the blood is proportional to the rate of infusion and inversely proportional to both the rate constant of elimination and the apparent volume of distribution.

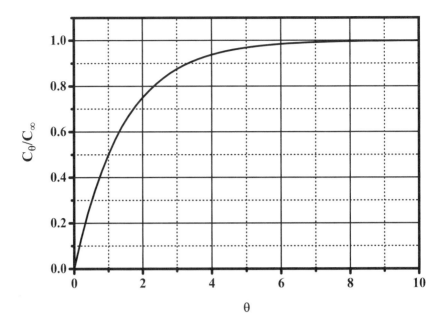

FIGURE 2.6 Master curve for the profile of blood concentration obtained during intravenous infusion at constant rate $\frac{C_\theta}{C_\infty} = f(\theta)$ with $\theta = \frac{t}{t_{0.5}}$ and $C_\infty = \frac{U}{k_e V_d}$.

Equation 2.24 can also be expressed in terms of the half-life time of the drug:

$$C_\theta = \frac{Ut_{0.5}}{V_p 0.693}[1-\exp(-0.693\theta)] = \frac{Ut_{0.5}}{V_p 0.693}[1-(0.5)^\theta] \tag{2.25}$$

For infinite time, or rather for a time long enough with regard to the half-life time, the term in exponential vanishes, and the asymptotical value of the concentration becomes

$$C_\infty = \frac{Ut_{0.5}}{V_p 0.693} = \frac{U}{V_p k_e} \tag{2.26}$$

As shown in Figure 2.6, a master curve is obtained by plotting the concentration at any time C_θ as a fraction of the asymptotical value of this concentration C_∞ in terms of the dimensionless time. This curve is of use whatever the drug and its pharmacokinetic parameters, such as the half-life time and the apparent volume of distribution. It is worth noting that the ratio of these concentrations is equal to 0.5 when $\theta = 1$, or in other words, when the time is equal to the half-life time of the drug.

The values given in Table 2.4 show that the concentration of the drug in the body increases with time to an asymptotic value, enabling the reader to know the time at which the process of the drug infusion may be considered performed under steady state.

TABLE 2.4
Time Necessary to Reach the Steady State (Constant Concentration)

θ	0	1	2	3	4	5	6	7	8	9	10	∞
$\dfrac{C_\theta}{C_\infty}$	0	0.5	0.75	0.87	0.94	0.97	0.984	0.992	0.996	0.998	0.999	1

The rate of administration U can thus be evaluated in terms of the half-life time of the drug, as well as by the apparent volume of distribution, in order to attain the optimal drug level in the patient's body, when these pharmacokinetic parameters are known.

2.3.2 CHANGE IN THE RATE OF DELIVERY DURING THE COURSE OF INFUSION

As mentioned, i.v. therapy exhibits some drawbacks in comparison with oral dosage forms. Nevertheless, another advantage of i.v. administration results from the fact that a change in the course of delivery is possible by varying the rate of infusion.

As already shown, with single doses or multidoses, the pharmacokinetic parameters vary greatly from one patient to another, and it should be necessary to adapt the rate of infusion to each patient. When neither the apparent volume of distribution nor the half-life time is known, it seems difficult to predict the right rate of drug adapted to a patient during the course of infusion. In fact, as shown in Figure 2.6 and more precisely in Table 2.4, it is not mandatory to deliver the drug at the first rate previously selected over a long period of time before making analyses, and thus for the therapist to decide to vary the rate of drug to adapt it to the patient.

The following protocol is proposed. The concentration of the free drug is measured at various times up to the time at which an insignificant change in the concentration occurs. From the values obtained, two are retained, let us call them C_1 and C_2, at times t_1 and t_2, respectively. The concentration C_1 is expressed in terms of time (or of θ) by Equation 2.25. At the longer time t_2, the concentration is nearly constant at the value given in Equation 2.26. The ratio of these two concentrations allows one to obtain the simple relation where only the time t_1 and the two concentrations appear:

$$\frac{C_1}{C_2} = 1 - (0.5)^{\theta_1} \tag{2.27}$$

with

$$\theta_1 = \frac{t_1}{t_{0.5}} \tag{2.28}$$

Thus, the unknown value of the half-life time is obtained:

$$t_{0.5} = t_1 0.693 \left[Ln \frac{C_2}{C_2 - C_1} \right]^{-1} \tag{2.29}$$

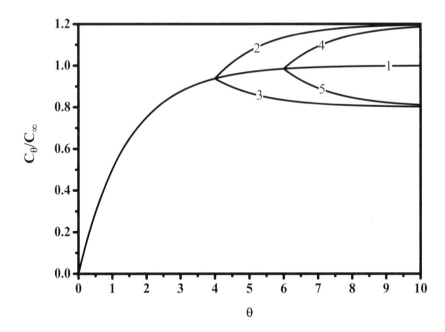

FIGURE 2.7 Master curve showing the profile of the free drug concentration obtained with infusion when the rate of infusion is varied by 20% of its initial value, at the dimensionless time $\theta = 4$ and $\theta = 6$.

The apparent volume of distribution is given by rewriting Equation 2.26:

$$V_p = \frac{Ut_{0.5}}{C_\infty 0.693} \tag{2.30}$$

Figure 2.7 shows the profiles of the drug concentration when the rate of infusion U is varied by 20% of its original value after the time t_2 used for calculating the pharmacokinetic parameters in the protocol.

2.3.3 VALUE OF THE AREA UNDER THE CURVE WITH INFUSION DELIVERY

The value of the AUC can be evaluated in the cases of infusion that follow. During CIV infusion up to the time t at which it is administered:

$$AUC_{0-t} = \frac{U}{k_e V_p} \int_0^t [1 - \exp(-k_e t)] dt = \frac{U}{k_e V_p} \left[t - \frac{1}{k_e} (1 - \exp(-k_e t)) \right] \tag{2.31}$$

This equation can be rewritten more simply when the time is long enough for the exponential term to vanish and the concentration to attain a constant value:

$$AUC_{0-t} = \frac{U}{k_e V_p}\left(t - \frac{1}{k_e}\right) = \frac{U t_{0.5}}{0.693 V_p}\left(t - \frac{t_{0.5}}{0.693}\right) \qquad (2.32)$$

When the intravenous infusion is stopped, the drug in the body is eliminated, leading to the following value of the AUC, which results from the elimination of the constant concentration of the drug from the body given in Equation 2.26:

$$AUC_{0-\infty} = \frac{U}{k_e V_p}\int_0^\infty \exp(-k_e t)dt = \frac{U}{V_p k_e^2} = \frac{U}{V_p}\left(\frac{t_{0.5}}{0.693}\right)^2 \qquad (2.33)$$

Of course, the total value of the AUC from the beginning to the end of the infusion process is obtained by adding the two values shown in Equation 2.32 and Equation 2.33.

2.4 REPEATED DOSES AT CONSTANT FLOW RATE OVER FINITE TIME

2.4.1 MATHEMATICAL TREATMENT OF THE PROCESS

For various reasons, hospital patients are given the drug dose intravenously at a constant flow rate U over a short period of time t_i (e.g., from 1 to 60 m); after this time, the plasma drug concentration is allowed to decrease over another period of time. After this sequence, the same process of drug delivery is repeated many times in succession (Figure 2.8).

The mathematical treatment can be drawn from both Equation 2.1 and Equation 2.3, as follows. During the time of infusion t_i, the plasma drug concentration is given by Equation 2.24, which can be rewritten in this way:

$$C_{ti} = \frac{Dose}{k_e V_d t_i}(1 - \exp(-k_e t_i)) \qquad (2.34)$$

In this case, the dose is expressed by

$$Dose = U t_i \qquad (2.35)$$

When the flow rate of drug is stopped at time t_i, the plasma drug concentration decreases until time T, according to Equation 2.2, becoming with respect to time:

$$C_t = C_{ti}\exp(-k_e(t - t_i)) \quad \text{with} \quad t_i < t < T \qquad (2.36)$$

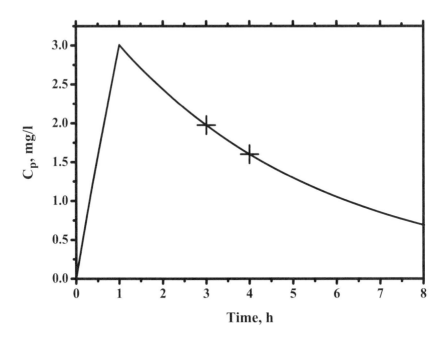

FIGURE 2.8 Plasma drug concentration vs. time with ciprofloxacin delivered at constant flow rate of 500 mg/h. At 3 h and 4 h, blood samples are taken for measurement, giving for the plasma drug concentration 1.97 and 1.60 mg/l, respectively, leading to $k_e = 0.21$/h and $V_p = 150$ l.

When it is desired to adapt the therapy to the patient, blood analyses are made at times t_1 and t_2 during this period of decreasing drug concentration. Calculation is made by following the method described in the Section 2.2.4, leading to the plasma drug concentrations:

$$C_{t1} = C_{ti} \exp(-k_e(t_1 - t_i)) \tag{2.37}$$

$$C_{t2} = C_{ti} \exp(-k_e(t_2 - t_i)) \tag{2.37'}$$

These two concentrations enable one to determine the pharmacokinetic parameters of the patient, whatever the protein binding:

$$k_e = \frac{1}{t_2 - t_1} Ln \frac{C_{t1}}{C_{t2}} \tag{2.3'}$$

$$C_{ti} = C_{t1} \exp(k_e(t_1 - t_i)) = C_{t2} \exp(k_e(t_2 - t_i)) \tag{2.4'}$$

$$V_d = \frac{Dose}{C_{ti}} = \frac{Ut_i}{C_{ti}} \tag{2.1'}$$

At time T, just before delivery of the second dose, the plasma drug concentration is

$$C_T = C_{ti} \exp(-k_e(T - t_i))$$ (2.38)

Thus, the plasma drug concentration after the second dose delivery, attained at time t_{2i}, is obtained using Equation 2.34, rewritten by taking into account the adequate values of the concentration and time:

$$C_{t2i} = \left(C_T + \frac{Dose}{k_e V_d t_i} \right)[1 - \exp(-k_e t_i)]$$ (2.39)

In fact, at time T, depending on the critical pharmacokinetic parameters of the patient found through Equation 2.3' and Equation 2.1', various parameters can be varied in order to adapt the therapy, such as the dose with the rate of drug delivery U and the time t_i over which it is delivered, as well as the time T, characterizing the frequency of the repeated doses.

2.4.2 GENERALIZATION OF THE PROCESSES

Equation 2.34 still holds for $t = 0$ in the case of the bolus injection. If it is true that for $t = 0$, this equation tends to an indeterminate form $\frac{0}{0}$, which has no value from the first approach, by applying Hôpital's rule, the ratio of the first derivatives is equal to $\frac{Dose}{V_d}$, and Equation 2.1 is obtained. Thus, Equation 2.34 is general for all the intravenous methods of drug delivery, either through bolus injection or through continuous infusion.

REFERENCES

1. Vidal Dictionnaire, Vidal, 2 rue Béranger, 75140 Paris, cedex 03, France.
2. Campoli-Richards, D.M. et al., Ciprofloxacin: a review of its antibacterial activity, pharmacokinetic properties and therapeutic use. Drug evaluation, *Drugs*, 35, 373, 1988.
3. Breilh, D. et al., Mixed pharmacokinetic population study and diffusion model to describe ciprofloxacin lung concentrations, *Comput. Biol. Med.*, 31, 147, 2001.
4. Wise, R. and Donovan, I., Tissue penetration and metabolism of ciprofloxacin, *Am J. Med.*, April 27, 103, 1987.
5. Fabre, D. et al., Steady-state pharmacokinetics of ciprofloxacin in plasma from patients with nosocomial pneumonia: penetration of the bronchial mucosa, *Antimicrob. Agents Chemother.*, Dec. 1991, 2521, 1991.
6. Smith, M.J. et al., Pharmacokinetics and sputum penetration of ciprofloxacin in patients with cystic fibrosis, *Antimicrob. Agents Chemother.*, Oct. 1986, 614, 1986.

3 Oral Dosage Forms with Immediate Release

NOMENCLATURE

a   Amount of drug initially in the dosage form (for calculation in Section 3.1.2).

AUC   Area under the curve expressing the drug concentration vs. time.

AUC_{0-t} Area under the curve from the time at which the dose is taken up to time *t*.

C   Free drug concentration in the blood.

$\frac{C_{max}}{C_\infty}$ Drug concentration at the peak in the blood as a fraction of C_∞.

C_∞ Drug concentration if all the drug in the dosage form were instantaneously in the blood.

GI(T)   Gastrointestinal (tract).

i.v.   Intravenous.

k_a Rate constant of absorption.

k_e Rate constant of elimination.

M   Amount of drug in various places (in Figure 3.2).

M_{in} Amount of drug initially in the dosage form (in Figure 3.2).

t, t_{max} Time, time at which the peak of concentration occurs, respectively.

T   Interval of time between doses.

$t_{0.5}$ Half-life time of the drug obtained with bolus i.v.

$T_{0.5}$ Half-life time of the drug obtained with oral dosage form.

$\theta = \frac{t}{t_{0.5}}$ Dimensionless time.

V_p Apparent volume of distribution.

w   Amount of drug eliminated from the blood.

y, z   Amount of drug remaining along the GIT, lying in the blood, respectively.

3.1 SINGLE DOSE

3.1.1 PRINCIPLE OF DRUG TRANSPORT

The scheme of drug transport clearly appears in the general diagram (Figure 3.1) showing the path followed by the drug from absorption into the organism to excretion from it. The stages are briefly described in succession:

- **Liberation** of the drug from the dosage form along the GIT occurs, either in the stomach, or in the intestine when a special coating protects the dosage form and the drug from the aggression of the acid gastric liquid. The supply form with its excipient exhibits a fast rate of dissolution, so as to provoke an immediate release of the drug.

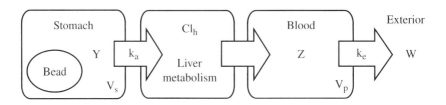

FIGURE 3.1 Path followed by the drug when taken orally.

- **Diffusion** of the drug takes place through the GI membrane into the liver and plasma compartment. In the liver, a first-pass hepatic may take place, giving active or inactive metabolites, this metabolic reaction follows a first-order kinetics in the case of linear pharmacokinetics. A main difference between intravenous delivery and oral delivery appears here; with oral delivery, the drug passes through the liver, where it can be metabolized.
- **Absorption** occurs, controlled by diffusion through the GI membrane and expressed in terms of rate transport by a first-order kinetics with the rate constant of absorption k_a.
- After binding to the plasmatic proteins, the stage of **distribution** into tissues of the free drug occurs, leading to the therapeutic action.
- Finally, the stage of **clearance and elimination** takes place, expressed by a first-order kinetics with the rate constant of elimination k_e.

Thus, three main pharmacokinetic parameters are necessary to describe the process:

- The rate constant of absorption
- The apparent volume of distribution (taking into account the amount of drug binding to the plasmatic proteins)
- The rate constant of elimination

The fundamental differences between the intravenous delivery and oral delivery of the drug lie in the presence of the diffusion of the drug through the GI membrane and the passage through the liver. As the rate constant of absorption is much larger than the rate constant of elimination, an accumulation of drug occurs in the blood, provoking the formation of a peak of concentration. As shown in Figure 3.2, where the amount of drug as a fraction of the initial amount of drug located in the dosage form is expressed in terms of time in the following sections, we observe along the GIT (curve 1) a strong decrease resulting from the instantaneous dissolution and fast rate of absorption, then an accumulation in the blood compartment with a peak of concentration (curve 2), and finally the amount of drug eliminated (curve 3). These three curves are obtained by calculation using ciprofloxacin as the drug.

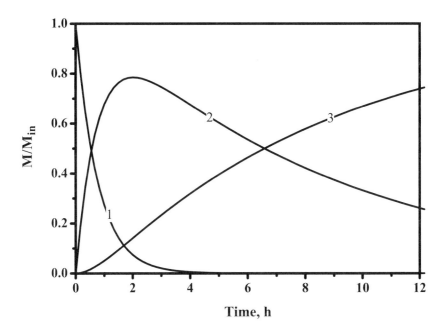

FIGURE 3.2 Kinetics of the drug in various compartments. 1. Amount of drug along the GIT. 2. Amount of drug in the blood. 3. Amount of drug eliminated.

3.1.2 CALCULATION OF THE PROFILES OF DRUG CONCENTRATION

By considering the amount of drug that has left the GIT at time t (noted y), the amount absorbed in the blood compartment (noted z), and finally the drug eliminated (called w), the three basic equations can be written to account for the successive two stages:

$$Y \xrightarrow{K_a} Z \xrightarrow{K_e} W$$

$t = 0$	a	0	0
t	$a - y$	z	w

$$\frac{dy}{dt} = k_a(a - y) \tag{3.1}$$

$$y = z + w \tag{3.2}$$

$$\frac{dw}{dt} = k_e z = k_e(y - w) \tag{3.3}$$

Equation 3.1, expressing the process of absorption, can be resolved independently, leading to the well-known equation:

$$y = a[1 - \exp(-k_a t)] \tag{3.4}$$

Equation 3.3, representing the amount of drug eliminated, becomes

$$\frac{dw}{dt} + k_e w = ak_e[1 - \exp(-k_a t)] \tag{3.5}$$

The treatment of the differential Equation 3.5 leads to the expression of the amount of drug eliminated w:

$$\frac{w}{a} = 1 + \frac{1}{k_a - k_e}[k_e \exp(-k_a t) - k_a \exp(-k_e t)] \tag{3.6}$$

Finally, by replacing w and y in Equation 3.2, the expression of the amount of drug in the blood z is given by

$$\frac{z}{a} = \frac{k_a}{k_a - k_e}[\exp(-k_e t) - \exp(-k_a t)] \tag{3.7}$$

The characteristics of the peak in the blood are obtained by differentiating Equation 3.7 with respect to time and writing that this new expression is equal to 0 (as the slope of the curve at the peak is flat). The two values characterizing the peak are the time at which it is attained t_{max} and the amount of drug present in the blood at that time z_{max} as a fraction of the initial amount of drug in the dosage form, a:

$$t_{max} = \frac{1}{k_a - k_e} Ln \frac{k_a}{k_e} \tag{3.8}$$

$$\frac{z_{max}}{a} = \left(\frac{k_e}{k_a}\right)^{\frac{k_e}{k_a - k_e}} \tag{3.9}$$

The concentration of the free drug in the blood is obtained by dividing the corresponding amount of drug by the apparent volume of distribution:

$$C = \frac{z}{V_p} \quad \text{and} \quad C_{max} = \frac{z_{max}}{V_p} \tag{3.10}$$

In Figure 3.2, curve 2, expressing the amount of drug in the blood as a fraction of the amount initially in the dosage form as a function of time represents in the same way the concentration of the free drug in the blood, because the pharmacokinetics is linear and Equation 3.10 holds. Ciprofloxacin is the drug, and for the values of the rate constant of absorption and elimination selected for calculation, the peak is attained at 2 h with a maximum concentration $\frac{C_{max}}{C_\infty}$ equal to 0.7848.

Let us note that C_∞ represents the concentration that would be obtained at time 0 for the bolus i.v. administration of a dose equal to the amount of drug, a, initially located in the dosage form. Thus, 2 h after the dosage form has been taken, the drug concentration at the peak is 0.7848 times the concentration that would be attained at time 0, C_0 (Chapter 2), with the bolus i.v. delivered with the same amount of drug.

The three stages of the overall process of drug delivery and transport through the body can be distinctly analyzed in this Figure 3.2. The stage of absorption remains active over 2 h after the uptake of the dosage form, up to the time at which the peak of concentration appears in the blood, and it is completely ended at less than 4 h. On the other hand, the stage of elimination starts after a little less than 1 h.

It is also worth noting that in Figure 3.2, curves 1 and 2, as well as curves 2 and 3, intersect at the ordinate 0.5. The reason for the interception of curves 1 and 2 is that the process of elimination does not interfere during the stage of absorption; curves 2 and 3 intercept because during the stage of elimination, the process of absorption has come to the end.

3.2   EFFECT OF THE PHARMACOKINETIC PARAMETERS' VALUES

The apparent volume of distribution plays a role only in decreasing the concentration of the free drug in the blood, so this parameter does not interfere in the ordinates of the various figures when the drug concentration in the blood is expressed as a fraction of the drug concentration C_∞. This value C_∞ is the drug concentration in the blood if the total amount of drug initially in the dosage form were instantaneously absorbed in the blood under the same conditions. C_∞ would be the same as that written C_0 in the case of bolus i.v. administration (Chapter 2, Section 2.1), by taking into account the effect of the first-pass hepatic.

The other pharmacokinetic parameters, such as the rate constants of absorption and elimination, act upon the rate of drug transport, so their effect can be seen not only along the time abscissa but also on the ordinate and especially on the value of the peak concentration.

3.2.1   EFFECT OF THE RATE CONSTANT OF ABSORPTION ON THE PROCESS

As already shown in Figure 3.2, the stage of absorption is very effective during the first 2 h of the process after the uptake of the dosage form. The effect of the value of the rate constant of absorption is precisely determined in Figure 3.3, where the ratio of the drug concentration $\frac{C}{C_\infty}$ is plotted against time, when the rate constant of absorption is varied between 1/h and 1.6/h, while the rate constant of elimination is kept constant at 0.12/h. The following three facts are worth noting:

- During the stage of absorption, the abrupt rise in the kinetics curve is sharper with the higher value of the rate constant of absorption (curve 3).
- The peak is higher and is reached at a shorter time when the rate constant of absorption is larger, as shown in curve 3.

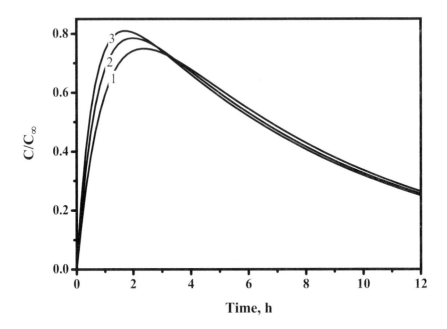

FIGURE 3.3 Effect of the value of the rate constant of absorption on the profiles of concentration of the drug (ciprofloxacin) in the blood as a function of time. $k_e = 0.12/h$; 1. $k_a = 1/h$. 2. $k_a = 1.3/h$. 3. $k_a = 1.6/h$.

- During the stage of elimination, the curve associated with the lower rate constant of absorption exhibits a longer value for the half-life time, but the difference between the extreme values of this half-life time does not exceed 0.5 h.

3.2.2 Effect of the Rate Constant of Elimination on Drug Transport

As shown in Table 2.1 (Chapter 2), the values of the rate constant of elimination vary greatly from one patient to another [Chapter 2, 1–6]. Calculation is made by considering the two extremes and the median values among those given in the literature for ciprofloxacin.

The profiles of the relative concentration $\frac{C}{C_\infty}$ of the drug in the blood are drawn in Figure 3.4 when the rate constant of absorption is 1.3/h and the rate constant of elimination is varied between 0.09/h and 0.22/h. These curves lead to the following conclusions:

- The three curves are well superimposed during the stage of absorption, before diverging around 0.5 h after the uptake.
- The characteristics of the three peaks are quite different, either for the time t_{max} or for the concentration $\frac{C_{max}}{C_\infty}$. The lower peak, reaching 0.7, is attained 1.6 h after the uptake for the higher value of the rate constant of

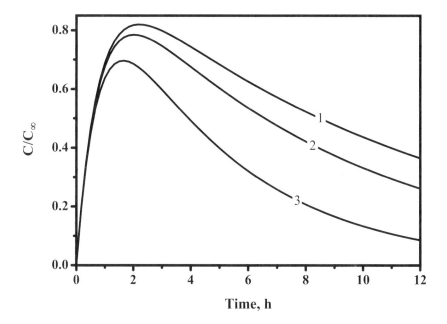

FIGURE 3.4 Effect of the value of the rate constant of elimination on the profiles of concentration of the drug (ciprofloxacin) in the blood as a function of time. $k_a = 1.3/h$; 1. $k_e = 0.09/h$. 2. $k_e = 0.12/h$. 3. $k_e = 0.22/h$.

elimination, whereas the higher peak at 0.82 is reached after 2.2 h for the lower value of this rate constant.
- Of greater concern is the half-life time of the drug obtained with these three values of the rate constant of elimination, since it varies from 10.86 h (curve 3 with $k_e = 0.22/h$) to 5.6 h (curve 1 with $k_e = 0.09/h$).

3.2.3 EFFECT OF THE NATURE OF THE DRUG

In this study concerning the nature of the drug, the effects of the rate constants of absorption and elimination are combined and they are thus considered together.

The profiles of concentration of the free drug in the blood are drawn in Figure 3.5 as they are calculated for various drugs:

1. Ciprofloxacin
2. Acetylsalicylic acid
3. Cimetidine

Some pharmacokinetics parameters found in the literature are collected in Table 3.1.

As shown in Table 3.1, the values of the pharmacokinetics parameters are scattered across a wide range. This fact results from two reasons:

- Different patients behave differently toward a drug.
- Ill patients and healthy volunteers respond to the drug in different ways.

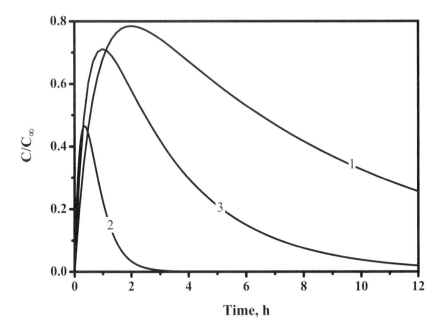

FIGURE 3.5 Effect of the nature of the drug on the profile of concentration in the blood as a function of time. 1. Ciprofloxacin with $k_a = 1.3/h$ and $k_e = 0.1155/h$. 2. Acetylsalicylic acid with $k_a = 3.5/h$ and $k_e = 2.1/h$. 3. Cimetidine with $k_a = 2.2/h$ and $k_e = 0.34/h$.

How patients and volunteers are selected is of great concern, as indicated by the great care taken in clinical trials:

- Phase 1, with 20 to 80 healthy volunteers
- Phase 2, taking care of the effectiveness over 100 to 300 patients
- Phase 3 especially, where effectiveness and the side-effect profile both were evaluated over 1000 to 3000 patients from many locations

Nevertheless, if the mean value for the pharmacokinetics parameters is perhaps very well defined, the dispersion may be larger. This variability of patient behavior toward the drug has already been considered in Chapter 2, with the values collected in Table 2.1, and displayed in Figure 3.3 and Figure 3.4 of this chapter.

TABLE 3.1
Pharmacokinetics Parameters of the Three Drugs

Drug	Half-life Time (h)	k_e(/h)	References
Ciprofloxacin	7.7–3.15	0.09–0.22	[1–10]
Acetylsalicylic acid	0.25–0.33	2.77–2.1	[11–13]
Cimetidine	1.2–3.3	0.57–0.21	[14–25]

TABLE 3.2
Characteristics of the Peaks of the Free Drug Concentration

Drug	k_e(h)	k_a(h)	t_{max}(h)	$\frac{C_{max}}{C_\infty}$	$T_{0.5}$(h)	$t_{0.5}$(h)
Ciprofloxacin	0.12	1.3	2	0.785	8.66	5.77
Acetylsalicylic acid	2.1	3.5	0.367	0.465	0.94	0.33
Cimetidine	0.34	2.2	1	0.711	3.6	2

Among the values presented in Table 3.1, some values for the half-life time are selected and used for calculation. These values are shown in Table 3.2, as well as the main characteristics obtained for the profiles drawn in Figure 3.5, such as the time at which the peak of concentration is attained and the maximum value reached for this concentration, as well as the half-life time obtained with this dosage form, which can be compared with the corresponding half-life time determined through bolus i.v. administration.

3.3 MULTIPLE ORAL DOSES

When oral doses are administered far apart in time, they behave independently. Because a minimum concentration is needed to maintain efficacy and a maximum concentration should not be exceeded to prevent side effects, the desired profile of concentration should be attained with repeated doses, especially for a long duration of therapy. Oral dosage forms must be taken sufficiently close together so that subsequent doses are administered in the blood prior to full elimination of the preceding dose. This system of drug delivery causes some accumulation of drug in the blood over a prolonged period of time, and the drug concentration alternates between peaks and troughs. Good knowledge of this process could be of help in optimizing the treatment to smooth out the side effects. Thus, a comparison can be made between oral dosage forms taken at regular intervals and repeated bolus i.v. injections. The ultimate objective, when these doses are taken at the right time intervals, would be to determine under which conditions the quality of the treatment may be considered equal to that obtained with i.v. injections.

3.3.1 PRINCIPLE FOR REPEATED ORAL DOSES AND PARAMETERS

The pharmacokinetics for the drug when taken orally is governed by three parameters, such as the rate constants of absorption and elimination and the apparent volume of distribution. Moreover, in multidose administration, the interval of time between successive doses becomes a parameter of great concern.

Calculation is made, using ciprofloxacin as the drug, by keeping constant the pharmacokinetic parameters of the drug and varying only the interval of time T within a wide range, such as 6 h, 12 h, 18 h.

The profiles of the drug concentration in the blood are depicted in Figure 3.6 with $T = 6$ h; in Figure 3.7 with $T = 12$ h; and in Figure 3.8 with $T = 18$ h. The values for

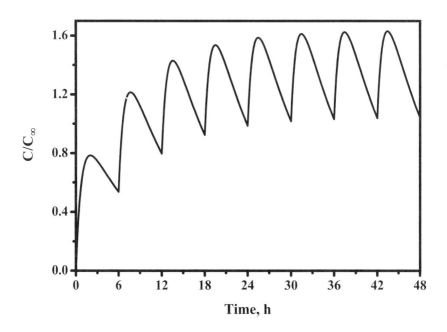

FIGURE 3.6 Profile of drug (ciprofloxacin) concentration with repeated doses, with the interval of time $T = 6$ h; $k_a = 1.3$/h and $k_e = 0.1155$/h; $t_{0.5} = 6$ h; $\theta = 1$.

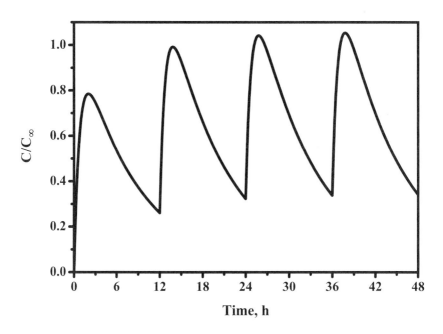

FIGURE 3.7 Profile of drug (ciprofloxacin) concentration with repeated doses, with the interval of time $T = 12$ h; $k_a = 1.3$/h and $k_e = 0.1155$/h; $t_{0.5} = 6$ h; $\theta = 2$.

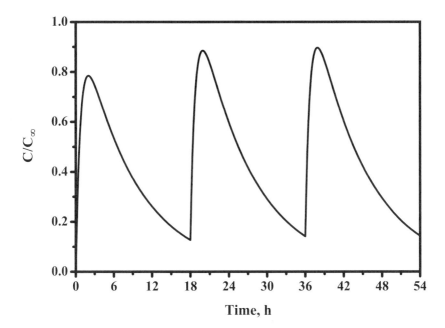

FIGURE 3.8 Profile of drug (ciprofloxacin) concentration with repeated doses, with the interval of time $T = 18$ h; $k_a = 1.3$/h and $k_e = 0.1155$/h; $t_{0.5} = 6$ h; $\theta = 3$.

the peaks and troughs are collected in Table 3.3, showing precisely the effect of the time interval between successive doses on the profile of the drug concentration in the blood.

From the observation of the three profiles of drug concentration depicted in Figure 3.6, Figure 3.7, and Figure 3.8 (and more particularly of the data gathered in Table 3.3), the following conclusions can be drawn:

- The interval of time between successive oral doses plays an important role, acting especially upon the values of the peaks and troughs.

TABLE 3.3
Effect of T on the Values of the Peaks and Troughs for Ciprofloxacin

T (h)/Dose	$k_a = 1.3$/h; $k_e = 0.12$/h; $t_{0.5} = 5.77$ h						
	1	2	3	4	6	7	8
6 Peak	0.785	1.21	1.43	1.53	1.59	1.61	1.62
Trough	0.537	0.80	0.92	0.98	1.02	1.03	1.04
12 Peak	0.785	0.99	1.04	1.05			
Trough	0.26	0.33	0.34				
18 Peak	0.785	0.885	0.897				
Trough	0.125	0.138	0.14				

- The time elapsed between two successive peaks slightly increases from the second dose to the next, finally reaching the value selected for the time interval between each dose. Thus, the time between two successive peaks is equal to the interval of time: after 6 doses for $T = 6$ h, but after 3 doses for $T = 12$ h, and after 2 doses for $T = 18$ h.
- The values of the drug concentration at the peaks and troughs increase from the first to the nth dose up to constant values, when the so-called steady state is attained. The steady state means that the same profile of concentration is reproduced for successive doses.
- The increase in the peaks and troughs depends on the value of the time interval between the successive doses. Table 3.3 shows that when the interval of time T is 6 h, under steady state, the concentrations at peaks and troughs are around twice the corresponding values attained for the first dose; this ratio is only 1.33 and 1.14 when the interval of time T is 12 h and 18 h, respectively.
- In conclusion, it clearly appears that the interval of time between two successive doses should be considered not only to facilitate the patient's compliance but also to obtain a smooth profile under steady state well situated between the effective and lethal levels of therapy, and thus to adapt the profiles of concentration within the frame between the *ED* and *LD* levels associated with the therapeutic index (shown in Chapter 1, Section 1.11).

3.3.2 EFFECT OF THE NATURE OF THE DRUG IN MULTIDOSES

Similarly to the case of a single dose, it is of great interest to consider the effect of the nature of the drug when it is taken in multidoses. The three drugs selected are the same as those studied with a single dose (i.e., ciprofloxacin, acetylsalicylic acid, and cimetidine); their pharmacokinetic parameters, such as the rate constant of absorption and elimination, are shown in Table 3.2. Thus, a comparison of the profiles of concentration obtained with a single dose and with multidose delivery should easily be possible. The shapes of the profiles of concentration are depicted in Figure 3.9, where the interval of time T is 12 h. Of course, the drug concentration alternates between high peaks and low troughs. The particular values measured for the successive peaks and troughs, from the first dose up to the steady state, are collected in Table 3.4 for these three drugs.

In Table 3.4, the maximum (peak) and minimum (trough) drug concentrations, $\frac{C_{max}}{C_\infty}$ and $\frac{C_{min}}{C_\infty}$, are noted Max and Min, respectively.

From the profiles drawn in Figure 3.9 and the data collected in Table 3.4, the following remarks are worth noting:

- As already shown in Figure 3.7 with ciprofloxacin, the profiles of concentration for the various drugs alternate between high peaks and low troughs.
- As already noted (Section 3.3.1), the time elapsed between two successive peaks is slightly lower than the interval of time T at the beginning of

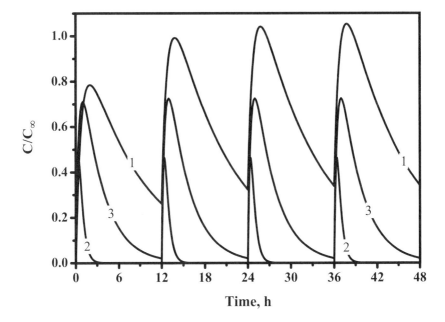

FIGURE 3.9 Profiles of drug concentrations with repeated doses for various drugs taken at the interval of time $T = 12$ h. 1. Ciprofloxacin. 2. Acetylsalicylic acid. 3. Cimetidine.

the process with multidoses. This time increases from dose to dose, up to the interval of time T, which is attained after three doses for ciprofloxacin.

- The values of the maximum concentrations at peaks and minimum concentrations at troughs increase from one dose to the next, up to the fourth dose, where the steady state is attained for ciprofloxacin. The steady state is already attained after the second dose for cimetidine. In the case of acetylsalicylic acid, there is no accumulation of drug because the interval of time T between two doses is much larger than the half-life time.

TABLE 3.4
Effect of the Nature of the Drug on the Concentrations at Peaks and Troughs

Drug	k_a(h)	k_e(h)	Dose	1	2	3	4
Ciprofloxacin	1.3	0.12	Max	0.785	0.992	1.042	1.053
			Min	0.26	0.325	0.337	0.34
Cimetidine	2.2	0.34	Max	0.71	0.725	0.725	
			Min	0.02	0.02		
Acetylsalicylic acid	3.5	2.1	Max	0.465	0.465		
			Min	0			

- The maximum concentration at peaks for a drug taken orally is lower than that attained with i.v. injections administered in multidoses, without considering the loss in concentration resulting from the first-pass hepatic. This reality is due largely to the fact that the process of absorption is not instantaneous (in contrast with bolus i.v. injection); also, the process of elimination already takes place during the stage of absorption.
- When the steady state is reached, the peaks and troughs are 1.34 times larger than those attained for the first dose in case of ciprofloxacin, whereas they are nearly identical for cimetidine. These constant values for cimetidine result from the fact that the interval of time T, taken at 12 h, is much larger than the half-life time of this drug, so that the next dose is delivered when the preceding dose has been eliminated. Thus, the value of the interval of time T of 12 h may be correct for ciprofloxacin, whose half-life time is 5.77 h, whereas it is too large for cimetidine, whose half-life time is 2 h. Obviously, the time interval between successive doses is far too long for acetylsalicylic acid, whose half-life time is one-third of an hour.
- From these observations, it can be said that the interval of time should be carefully selected in relation to the value of the half-life time of the drug, for a prolonged therapy necessitating a steady profile of concentration alternating between peaks and troughs not too far apart, and totally situated within the optimal therapeutic level.

3.3.3 MASTER CURVES WITH DIMENSIONLESS NUMBER

The best way to find an easy solution is to build and use master curves based on a dimensionless number expressing the time. In master curves, the concentration of the drug at time t, as a fraction of the concentration associated with the total amount of drug initially in the GIT, is expressed in terms of the *dimensionless time*, which is the time as a fraction of the half-life time of each drug. Thus, this master curve can be used whatever the drug and its half-life time. Two values of the interval of time between doses are selected:

- When the time interval is equal to the half-life time of each drug with $\theta = 1$ (Figure 3.10)
- When it is twice their half-life times ($\theta = 2$) (Figure 3.11)

The protocol for using these master curves is as follows. By considering a drug whose half-life time is known (e.g., ciprofloxacin with $t_{0.5} = 6$ h), when the interval of time between successive doses is $T = 6$ h, there is $\theta = 1$. Then the profile of concentration is developed from dose to dose, according to curve 1 shown in Figure 3.10. On the other hand, when the interval of time T is 12 h, with $\theta = 2$, the profile of concentration for the same drug is drawn in curve 1 in Figure 3.11.

The same protocol can be followed for any drug (e.g., with cimetidine, for which $t_{0.5} = 2$ h) by selecting the value of 2 h for the interval of time between doses and by looking at the profile shown in curve 2 in Figure 3.10, or the value of 4 h for the interval of time by considering curve 2 in Figure 3.11.

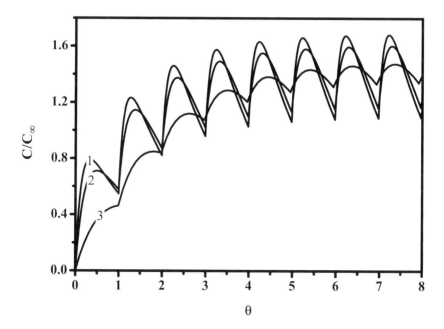

FIGURE 3.10 Master curves drawn for the profiles of concentration of drugs when the doses are taken in succession with the interval of time T equal to the half-life time of each drug ($\theta = 1$). 1. Ciprofloxacin with $k_a = 1.3/h$, $k_e = 0.1155/h$, and $t_{0.5} = 6$ h. 2. Cimetidine with $k_a = 2.2/h$ $k_e = 0.34/h$, and $t_{0.5} = 2$ h. 3. Acetylsalicylic acid with $k_a = 3.5/h$, $k_e = 2.1/h$, and $t_{0.5} = 0.33$ h.

In the particular case of acetylsalicylic acid, for which the half-life time is as short as 0.33 h, the interval of time T is also 0.33 h, so as to have $\theta = 1$ for curve 3 in Figure 3.10, and it is 0.66 h with $\theta = 2$ in Figure 3.11.

The following remarks and conclusions can be made:

- A steady state is attained, with steady values for the peaks and troughs, after a few doses, for example, around five to seven doses when the interval of time is equal to the half-life time ($\theta = 1$) and around three to four doses when the interval of time is twice the half-life time ($\theta = 2$).
- The values of the maximum (peak) and minimum (trough) concentrations increase from dose to dose, reaching the values found under the steady state. When the period of time is equal to the half-life time ($\theta = 1$), the concentrations at peaks and troughs under steady state are about twice those obtained in the first dose for ciprofloxacin and cimetidine. With a period of time twice the half-life time ($\theta = 2$), the increase in the concentration is less important, and the concentrations at peaks and troughs under steady state are 1.33 to 1.39 times larger than those obtained in the first dose, for the same two drugs.
- Under steady-state conditions, the ratio of the maximum to the minimum concentration is around 1.6 when $\theta = 1$ and 3 when $\theta = 2$ for ciprofloxacin, and 1.4 and 2.5, respectively, for cimetidine. In the particular case of

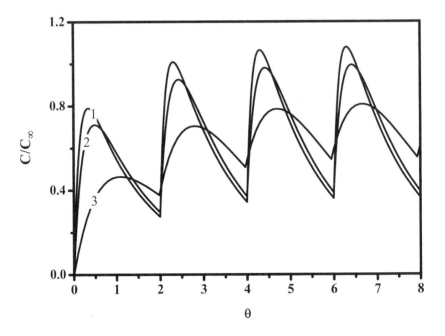

FIGURE 3.11 Master curves drawn for the profiles of concentration of drugs when the doses are taken in succession with the interval of time T equal to twice the half-life time of each drug ($\theta = 2$). 1. Ciprofloxacin with $k_a = 1.3$/h, $k_e = 0.1155$/h, and $t_{0.5} = 6$ h. 2. Cimetidine with $k_a = 2.2$/h, $k_e = 0.34$/h, and $t_{0.5} = 2$ h. 3. Acetylsalicylic acid with $k_a = 3.5$/h, $k_e = 2.1$/h, and $t_{0.5} = 0.33$ h.

acetylsalicylic acid, the profile of concentration is quite different from the other two drugs, as shown in Figure 3.10, curve 3, resulting from the fact that the subsequent dose is taken before the peak of the preceding one has been reached.

- Thus, an increase in the interval of time between doses, T, with respect of the half-life time (e.g., θ, from 1 to 2) is responsible for two facts:
  - The steady state is attained after the same time.
  - The ratio of the maximum concentration to the minimum concentration is increased from 1.6 to 3 for ciprofloxacin and from 1.4 to 2.5 for cimetidine.
- Note that the increase in the maximum and minimum concentrations with bolus i.v. administration is twice that obtained in the first dose when the interval of time between two injections is equal to the half-life time of the drug (Chapter 2, Section 2.2.1). Thus, a comparison of oral dosage forms with immediate release and bolus i.v. shows that, with these two systems of drug administration, multidoses lead to the same increase in the maximum and minimum concentrations. The difference comes from the fact that the concentration is always lower with oral dosage forms than with bolus i.v., without considering the decrease resulting from the metabolic reaction in the liver due to the first-pass hepatic.

3.4 AREA UNDER THE CURVE WITH ORAL DOSAGE FORMS

The area under the curve (AUC) can be calculated either with a single dose or with repeated doses. In this calculation, the effect of the first-pass hepatic is not considered. However, the metabolism following this first-pass hepatic can be taken into account, provided that some data were supplied from in vivo measurements, either with bolus i.v. or with an oral dosage form and the same amount of drug delivered.

3.4.1 AUC WITH A SINGLE DOSE

The profile of the free drug concentration in the blood is expressed by Equation 3.7 rewritten by using the ratio of the concentrations:

$$\frac{C_t}{C_\infty} = \frac{k_a}{k_a - k_e}[\exp(-k_e t) - \exp(-k_a t)] \tag{3.11}$$

where C_t is the concentration at time t, and C_∞ the concentration associated with the amount of drug in the dosage form (which is equal to the concentration C_0 obtained with bolus i.v. by neglecting the effect of the first-pass hepatic).

By integrating the preceding equation with respect to time between the limits 0 and ∞, the following general equation is obtained:

$$AUC_{0-\infty} = \int_0^\infty C_t dt = \frac{C_\infty k_a}{k_a - k_e}\left[\frac{1}{k_e}\exp(-k_e t) - \frac{1}{k_a}\exp(-k_a t)\right]_\infty^0 \tag{3.12}$$

which reduces to

$$AUC_{0-\infty} = \frac{C_\infty}{k_e} = \frac{C_\infty t_{0.5}}{0.693} \tag{3.13}$$

Note that this value of the AUC associated with a single dose is equal to the value obtained by integrating Equation 2.2 under the same conditions of time with bolus i.v. Thus, the AUC calculated in the case of a single dose taken orally is equal to that obtained with a bolus i.v., without considering the first-pass hepatic, because the metabolism can reduce the amount of drug available in the case of oral delivery. Thus, according to the value of the bioavailability found with ciprofloxacin varying within the range of 45 to 77%, the actual value of the AUC associated with a single oral dose will be 45 to 77% that obtained with the bolus i.v.

3.4.2 AUC WITH REPEATED DOSES

A dimensionless number is introduced for expressing the results by dividing the AUC associated with each dose taken in succession with the same time interval by the AUC obtained with a single dose far apart from the others. This ratio is written as follows:

$$\frac{AUC_{nT-(n+1)T}}{AUC_{0-\infty}} = k_e AUC_{nT-(n+1)T} = \frac{0.693}{t_{0.5}} AUC_{nT-(n+1)T} \tag{3.14}$$

TABLE 3.5
Values of $AUC_{nT-(n+1)T}$ as a Fraction of $AUC_{0-\infty}$ for Each Drug

Drug/	Doses	1	2	3	4	5	6	7	8
1	$\theta = 1$	0.45	0.72	0.86	0.93	0.97	0.98	0.99	0.996
Curve 1	$\theta = 2$		0.72		0.93		0.98		0.996
2	$\theta = 1$	0.41	0.70	0.85	0.93	0.96	0.98	0.99	0.995
Curve 2	$\theta = 2$		0.70		0.93		0.98		0.99
3	$\theta = 1$	0.22	0.52	0.73	0.86	0.93	0.96	0.98	0.99
Curve 3	$\theta = 2$		0.52		0.86		0.96		0.99

where $AUC_{0-\infty}$ is the area under the curve associated with a single dose and $AUC_{nT-(n+1)T}$ represents the area under the curve between the time nT and $(n+1)T$, associated with the nth dose when these doses are taken in succession with the constant interval of time T, and k_e is the rate constant of elimination of the drug.

The values of this ratio of the area under the curve are presented in Table 3.5 for the following three drugs, when they are taken in succession with the interval of time either $\theta = 1$ or $\theta = 2$:

1. Ciprofloxacin
2. Cimetidine
3. Acetylsalicylic acid

The profiles of drug concentration obtained with the doses taken in succession, shown in Figure 3.10 and Figure 3.11, and the values of the AUC collected in Table 3.5, enable one to draw the following conclusions:

- The concentration increases from dose to dose, exhibiting peaks and troughs (as already shown in Table 3.3 for ciprofloxacin and Table 3.4 for the same three drugs), whatever the interval of time elapsed between each dose uptake. After a number of doses, depending on the value of the interval of time (e.g., 5 to 7 for $\theta = 1$ and 3 to 4 for $\theta = 2$), the stationary state is attained. The AUC associated with each dose between each interval of time increases in the same way, up to a constant value when the stationary state is attained.
- Under stationary-state conditions, each dose is responsible for an AUC equal to that obtained with a single dose taken alone far apart from the others.
- The value of the time interval between successive doses acts upon the profiles of concentration, as well as upon the value of the AUC. It is worth noting that the AUC associated with each even-numbered dose taken with $\theta = 1$ is nearly the same as that obtained with $\theta = 2$. Thus, when the steady state is attained, the AUC is proportional to the number of doses taken, whatever the interval of time between the dose uptakes.

3.5   COMPARISON OF ORAL AND I.V. DRUG DELIVERY

The *dosage regimen* or *schedule* is the plan of administration of one or several drugs. The dosage regimen is prescribed by the physician, who bases the decision on the information presented by the drug manufacturer. An exact dosage schedule is of particular importance when it is necessary to maintain a constant therapeutic concentration in the blood and tissues over a long period of time.

For very serious diseases, it is best for the patient to have the drug administered at the hospital. There, the drug can be delivered either through repeated i.v. injections, with the interval of time precisely defined depending on the patient's pharmacokinetics and pharmacodynamics response, or through continuous infusion of the right dose.

For less serious diseases, the therapy can be accomplished at home by the patient him- or herself; the problem becomes the patient's compliance. Almost all conventional self-administered drugs through oral dosage forms first reach a therapeutic peak concentration and then, because of their short biological half-life time, rapidly fall below the minimum concentration for therapeutic activity. At this point, patient compliance and the limited reliability of the patient intervene. Often the physician is unaware of how frequently the patient changes the carefully prescribed dosage schedule. Investigations conducted in the United States and Europe demonstrated that at least 50% of patients do not comply with the regimen. This percentage varies from disease to disease, and it can be said that patients have more difficulty in complying with the treatment of a disease causing no apparent complaint.

Therapeutic systems, and especially oral dosage forms with controlled release, certainly offer the opportunity to improve patient compliance by reducing some problems connected with their immediate-release counterparts. All compliance studies make it clear that reducing the frequency of drug administration improves compliance, and a once-a-day (or at most twice-a-day) schedule is of help in establishing a routine procedure, with the uptake occurring either in the morning when the patient is getting out of bed or in the evening just before the patient gets into bed. Thus, oral controlled-release has been a big winner for the drug delivery industry [26]. More important, not only are these dosage forms a commodity item in improving patient compliance, but also patients and physicians both find value in simplified dosing and reducing side effects, which improve ease of use and the quality of life. Finally, consumers are concerned with improved compliance and decreased health care costs.

REFERENCES

1. Ball, A.P. et al., Pharmacokinetics of oral ciprofloxacin, 100 mg single dose, in volunteers and elderly patients, *J. Antimicrob. Chemother.,* 17, 629, 1986.
2. Terp, D.K. and Rybak, M.J., Ciprofloxacin, *Drug Intell. Clin. Pharm.,* 21, 568, 1987.
3. Nix, D.E. et al., Effect of multiple dose oral ciprofloxacin on the pharmacokinetics of theophilline and indocyanine green, *J. Antimicrob. Chemother.,* 19, 263, 1987.
4. Parry, M.F., Smego, D.A., and Digiovanni, M.A., Hepatobiliary kinetics and excretion of ciprofloxacin, *Antimicrob. Agents Chemother.,* 7, 982, 1988.

5. Bergan, T., Pharmacokinetics of ciprofloxacin. Documentation Bayer. Pharma, *Excerpta Medica,* 45, 1988.

6. Brogard, J.M. et al., Place des métabolites de la ciprofloxacine dans son élimination biliaire et urinaire chez l'homme, *Pathologie Biologie,* 36, 5 bis, 719, 1988.

7. Campoli-Richards, D.M. et al., Ciprofloxacin: a review of its antibacterial activity, pharmacokinetic property and therapeutic use. Drug evaluation, *Drugs,* 35, 373, 1988.

8. Ludwig, E. et al., The effect of ciprofloxacin on antipyrine metabolism, *J. Antimicrob. Chemother.,* 22, 61, 1988.

9. Davey, P.G., Overview of drug interactions with the quinolones, *J. Antimicrob. Chemother.,* 22, suppl. C, 97, 1988.

10. Breilh, D. et al., Mixed pharmacokinetic population study and diffusion model to describe ciprofloxacin lung concentrations, *Comput. Biol. Med.,* 31, 147, 2001.

11. Clissold, S.P., Aspirin and related derivatives of salicylic acid, *Drugs,* 32, suppl. 4, 8, 1986.

12. Verbeeck, R.K., Blackburn, J.L., and Loewen, G.R., Clinical pharmacokinetics of non-steroidal anti-inflammatory drugs, *Clin. Pharmacokin.,* 8, 297, 1983.

13. Vidal Dictionnaire, Vidal, 2 rue Béranger, 75140, Paris, cedex 03, France.

14. Somogyi, A. and Roland, G., Clinical pharmacokinetics of cimetidine, *Clin. Pharmacokin.,* 8, 463, 1983.

15. Arancibia, A. et al., Pharmacocinétique de la cimetidine après administration d'une dose unique par voie i.v. rapide et d'une autre voie orale, *Thérapie,* 40, 87, 1985.

16. Mihaly, G.W. et al., Pharmacokinetic studies of cimetidine and ranitidine before and after treatment in peptic ulcer patients, *Br. J. Clin. Pharmacol.,* 17, 109, 1984.

17. Oberle, R.L. and Amidon, G.L., The influence of variable gastric emptying and intestinal transit rates on the plasma level curve of cimetidine; an explanation for the double peak phenomenon, *J. Pharmacokin. Biopharm.,* 15, 529, 1987.

18. Bodemar, G. et al., The absorption of cimetidine before and during maintenance treatment with cimetidine and the influence of a meal on the absorption of cimetidine. Studies in patients with peptic ulcer diseases, *Br. J. Clin. Pharmacol.,* 7, 23, 1979.

19. Vincon, G. et al., Influence de l'alimentation sur la biodisponibilité de la cimetidine, *Thérapie,* 38, 607, 1983.

20. Nation, R.L. et al., Pharmacokinetics of cimetidine in critically ill patients, *Eur. J. Clin. Pharmacol.,* 26, 341, 1984.

21. Gonzales-Martin, G. et al., Pharmacokinetics of cimetidine in patients with liver disease, *Int. J. Clin. Pharmacol.,* 23, 355, 1985.

22. Bauer, L.A. et al., Cimetidine clearance in the obese, *Clin. Pharmacol. Ther.,* 37, 425, 1985.

23. D'Angio, R. et al., Cimetidine absorption during sucralfate, *Br. J. Clin. Pharmacol.,* 21, 515, 1986.

24. Keller, E. et al., Increased nonrenal clearance of cimetidine during antituberculous therapy, *Int. J. Clin. Pharmacol.,* 22, 307, 1984.

25. Kirch, W. et al., Influence of β-receptor antagonists on pharmacokinetics of cimetidine, *Drugs,* 25, 127, 1983.

26. Henry, C.M., New wrinkles in drug delivery, *Chem. Eng. News,* 37, March 1, 2004.

4 Kinetics of Drug Release from Oral Sustained Dosage Forms

NOMENCLATURE

a, b, c Half the sides of a solid, parallelepiped, cube, sheet.

β_n Positive roots of $\beta \tan \beta = G$ in Equation 4.8, for a sheet.

C, C_s, C_{ext} Concentration, on the surface, in the external liquid.

D Diffusivity, or coefficient of diffusion (cm²/s).

$\frac{\partial C}{\partial x}, \frac{\partial C}{\partial r}$ Longitudinal, radial gradient of concentration, respectively.

F Flux of diffusing substance mass/unit area/unit time.

G Dimensionless number $G = \frac{hL}{D}$ in Equation 4.9, for a sheet.

h Coefficient of convection on the surface of the dosage form (cm/s).

H Half the length of a cylinder.

γ_v Roots of Equation 4.25, for a sphere.

K Dimensionless number in Equation 4.24, for a sphere.

L Half the thickness of a sheet.

M_t Amount of drug released up to time t.

M_∞ Amount of drug released after infinite time.

M_{in} Amount of drug initially in the dosage form.

n, m, p Integers used for calculation in series.

q_n Roots of the Bessel function in radial diffusion, in Equation 4.33 for a cylinder.

R Radius of the sphere, of a cylinder.

t, t_r Time, time of full erosion of the dosage form, respectively.

v Linear rate of erosion (Equation 4.38).

x, r Longitudinal, radial abscissa along which diffusion takes place, respectively.

DBS, ATBC plasticizer Dibutyl sebacate, acetyl tributyl citrate, respectively.

The various sustained-release oral dosage forms are generally made of polymers through which the drug is dispersed. Depending on the nature of the polymer, the process of drug release is controlled either by diffusion or by erosion, and very often by both these processes. Thus we will consider these three processes in succession.

Experimentally, this kinetics of drug release is determined through in vitro measurements. Nevertheless, calculation is always useful, showing the best line to be taken for the research, allowing considerable reduction of the number of tedious experiments and offering logical shortcuts. Moreover, the equations expressing the kinetics of drug release are necessary for calculating the plasma drug level.

4.1 DRUG RELEASE CONTROLLED BY DIFFUSION

Generally in this case, the polymer is stable in the GI and passes unchanged through the body. It is not the purpose of this book to examine deeply the process of diffusion through polymers. Three books are worth noting for readers interested in such a study. The basic book is *The Mathematics of Diffusion* [1], in which Crank successfully applied to diffusion the results obtained in a fundamental study on heat conduction [2]. In this book, the mathematical treatment is made by selecting the various shapes that we will consider in succession—sheets and parallelepipeds, spheres and cylinders—when the diffusivity is constant. In the second book, both mathematical and numerical treatments of diffusion are made, allowing the reader to consider more complex matter transfers—especially when the diffusivity is concentration dependent—and various practical applications are presented [3]. The third book is devoted to the applications of diffusion for drug liberation from oral sustained-release dosage forms, as they are determined through in vitro tests [4].

4.1.1 PRINCIPLES OF DIFFUSION AND BASIC EQUATIONS

Diffusion is the process through which matter is transported from one place to another, resulting from random molecular motions. On average, the matter is transferred from the region of higher concentration of matter to that of lower concentration. The first typical example is that of Democritus (4th century B.C.), who dropped resinous wine into motionless water. The diffusion led him to consider the small particles of this dye and predict the existence of atoms.

The mathematical equation of heat conduction was established by Fourier in 1822. A few decades later, in 1855, Fick put diffusion on a quantitative basis by adopting the same equation. In an isotropic substance, the rate of diffusion through unit area of a section is proportional to the gradient of concentration measured normal to this section:

$$F = -D\frac{\partial C}{\partial x} \tag{4.1}$$

where D is the diffusivity and $\frac{\partial C}{\partial x}$ is the gradient of concentration, the negative sign arising because diffusion occurs in the opposite direction to that of increasing concentration.

4.1.1.1 Boundary Conditions

The *boundary conditions* express the value of the concentration and of the gradient of concentration of diffusing substance on the surface of the dosage form.

When the drug does not diffuse out of the dosage form, the mathematical condition is obtained by writing that the rate of transfer through its surface is 0:

$$\frac{\partial C}{\partial x} = 0 \quad \text{on the surface} \tag{4.2}$$

When the rate of matter transfer through the external surface is finite, a finite coefficient of matter transfer through this surface is introduced, leading to the following general equation:

$$-D\left(\frac{\partial C}{\partial x}\right)_s = h(C_s - C_{ext})$$ (4.3)

where C_s and C_{ext} are the drug concentration on the surface of the dosage form and in the surrounding liquid, respectively.

This condition on the surface expresses that the rate at which the drug leaves the surface is constantly equal to the rate at which the drug is brought to this surface by internal diffusion. This rate, per unit area, is proportional to the difference between the actual concentration on the surface C_s and the concentration required to maintain equilibrium with the surrounding liquid C_{ext}.

When the coefficient of matter transfer is very large, tending to infinity, it clearly appears that the concentration on the surface is equal to that in the surrounding liquid as soon as the dosage form is put in contact with the liquid and the process of diffusion starts:

$$C_s = C_{ext} \quad \text{at any time and especially at } t = 0$$ (4.4)

4.1.1.2 Initial Conditions

Generally, the drug concentration is initially uniform through the dosage form, meaning that there is no gradient of drug concentration in it.

4.1.1.3 Process of Drug Release

It clearly appears that the drug is only released from the dosage form when it is brought into contact with a liquid. The process of drug release is thus far more complex, as was proved in various studies, one with Eudragit and another with Carbopol [5, 6], when dosage forms made of these polymers and acetylsalicylic acid as the drug were immersed in synthetic gastric liquid, following a previous study performed on plasticized PVC [7]. A double matter transfer takes place, as the liquid diffuses through the polymer and dissolves the drug, enabling the drug to diffuse through the liquid located in the polymer. Thus, it is easy to understand that the diffusivity of the drug depends on the local concentration of liquid situated in the polymer. The solution to this difficult problem necessitates complex experiments, weighing the dosage form and measuring the drug concentration in the liquid simultaneously, as to determine the kinetics of the transfer of both the liquid and the drug, and a numerical treatment is necessary for calculation [3, 4]. Because pharmacists are only interested in the drug release in the liquid, however, only the diffusion of the drug alone through the polymer will be considered in the following chapters.

4.1.1.4 In Vitro Dissolution Tests

Biopharmaceutical characterization of drug formulations requires a suitable dissolution model apparatus. Most experience gained over the years of in vitro dissolution
testing has been with models designed to be applied to sustained-release dosage
forms. The paddle and the rotating basket models are described in USP XXII,
European, Japanese, and other pharmacopoeias worldwide. The alternative flow-
through cell was also developed if a change in pH in the test medium is required [8].

In order to increase their predictive value, dissolution tests aim to mimic in vivo
behavior along the GIT, so the models are adjusted to simulate the physiological
conditions as closely as possible [9]. From numerous scientific data, it has been
concluded that there is no need for too-close simulation of physiological conditions.
However, special care with pH and agitation with the stirring rate should give
conditions similar to physiological reality and so avoid artificial data. It is said that
high rates of agitation are not necessary [10]; nevertheless, it is true that in motionless
liquid the process of drug release is controlled not only by diffusion through the
dosage form but also by convection in the liquid, and increasing the stirring rate
reduces the convection effect. This fact is especially considered in the following
sections, after the parameters of diffusion have been determined.

4.1.1.5 Parameters of Diffusion

Finally, if the diffusivity D remains the essential parameter of diffusion, the coefficient of convection h at the liquid–dosage form interface may appear of concern,
especially at the beginning of the process when the stirring rate is too low. The shape
given to the dosage form, and more particularly to its dimensions, represents the
other parameters of interest.

4.1.2 Kinetics of Drug Release from Thin Films (Thickness 2 *L*)

Fickian diffusion is expressed by the second law of diffusion (which can be established from Equation 4.1, called the first law of diffusion, by evaluating the matter
balance through a thin sheet of thickness dx during the small time dt). This can be
written in the form of a partial derivative equation:

$$\frac{\partial C}{\partial t} = D\frac{\partial^2 C}{\partial x^2} \tag{4.5}$$

where C is the concentration of the drug in the dosage form, depending on time t
and position x, and D is the diffusivity of the drug through the dosage form.

The respective initial and boundary conditions are written as follows:

$$t = 0 \quad -L \le x \le +L \quad C = C_{in} \tag{4.6}$$

$$t > 0 \quad -D\left(\frac{\partial C}{\partial x}\right)_L = h(C_s - C_{ext}) \tag{4.3}$$

At the beginning of the experiment ($t = 0$), the drug is uniformly distributed through the film of thickness 2 L, C_{in} being this concentration.

Equation 4.5 has two types of solutions: one expressed in terms of an exponential series, and the other in terms of error-function series. Three calculation methods exist: separation of variables, the Laplace transform, and the method based on superposition and reflection in a finite system [1–4].

The general solution of the problem set up by Equation 4.5 and Equation 4.3 with a finite coefficient of convection h at the interface is given by [1–4]:

$$\frac{M_\infty - M_t}{M_\infty} = \sum_{n=1}^{\infty} \frac{2G^2}{\beta_n^2(\beta_n^2 + G^2 + G)} \exp\left(-\frac{\beta_n^2}{L^2} Dt\right) \tag{4.7}$$

where the β_n are the positive roots of

$$\beta \tan \beta = G \tag{4.8}$$

and the dimensionless term G is expressed by

$$G = \frac{hL}{D} \tag{4.9}$$

where L is half the thickness of the sheet.

Some roots of Equation 4.8 are supplied for various values of G in the tables [1–4], but the best way is to calculate them, because a table cannot provide all the β_n values for any value of G.

When the rate of stirring is high enough, the coefficient h becomes so large that it can be considered infinite, and the drug concentration on the surface C_s reaches its value at equilibrium as soon as the dosage form is put in contact with the liquid. In this case, the β_n are expressed in terms of π and Equation 4.7, when $G > 50$–100,

TABLE 4.1
Diffusion Parameters of Drugs in Ethyl Cellulose with Various Plasticizers

Drug	%w/w	Plasticizer	Thickness (μm)	$D\,10^{-10}\,(cm^2/s)$	$h\,10^{-6}\,(cm/s)$
Theophylline	ATBC	20	55	5.0 (±1.8%)	18 (±6%)
Diltiazem HCl	DBS	10	173	2.5 (±3.6%)	3.2 (±3%)
		15	173	3.4 (±4.7%)	4.0 (±2.3%)
		20	173	5.7 (±10%)	4 (±4%)
		25	173	13 (±3%)	14 (±2%)
Caffeine	ATBC	20	331	8.2 (±4%)	5 (±2%)

Reprinted From J. Phermac. Sci. 87(7), pp. 827–832, 1998. With permission from Wiley-Liss, Inc., a Wiley company.

reduces to

$$\frac{M_\infty - M_t}{M_\infty} = \frac{8}{\pi^2} \sum_{n=0}^{\infty} \frac{1}{(2n+1)^2} \exp\left[-\frac{(2n+1)^2 \pi^2}{4L^2} Dt\right] \qquad (4.10)$$

where M_t and M_∞ are the amounts of drug released after time t and infinite time, respectively. In fact, M_∞ is also the amount of drug initially located in the dosage form, when the volume of liquid is more than 50 times that of the dosage form, n is an integer ranging from 1 to infinity (generally 5 to 7 is enough), and L is half the thickness of the sheet.

By using the error function, the following series is obtained [1–4]:

$$\frac{M_t}{M_\infty} = \frac{2\sqrt{Dt}}{L}\left[\frac{1}{\sqrt{\pi}} + 2\sum_{n=1}^{\infty} (-1)^n ierfc\left(\frac{nL}{\sqrt{Dt}}\right)\right] \qquad (4.11)$$

where *ierfc* is the integral of the error-function complement, and n is an integer.

Equation 4.10 and Equation 4.11 reduce significantly, either for long times or for short times, this time being related to the ratio $\frac{M_t}{M_\infty}$:

$$t \text{ is large with } \quad \frac{M_t}{M_\infty} > 0.5-0.7 \quad \frac{M_\infty - M_t}{M_\infty} = \frac{8}{\pi^2} \exp\left[-\frac{\pi^2 Dt}{4L^2}\right] \qquad (4.12)$$

$$t \text{ is small with } \quad \frac{M_t}{M_\infty} < 0.6 \quad \frac{M_t}{M_\infty} = \frac{2}{L}\sqrt{\frac{Dt}{\pi}} \qquad (4.13)$$

Equation 4.12 and Equation 4.13 lead to the following simple relation:

$$\text{For} \quad \frac{M_t}{M_\infty} = 0.5 \quad \frac{Dt}{L^2} = 0.196 \qquad (4.14)$$

4.1.2.1 Dimensionless Numbers

Two dimensionless numbers are of great concern: one related to the kinetics, expressed in terms of time, and the other related to the coefficient of convection at the interface:

$$\frac{Dt}{L^2} \quad \text{and} \quad G = \frac{hL}{D} \qquad (4.15)$$

The first one shows that, for a given diffusivity associated with the drug–polymer couple, the time of release is proportional to the square of the thickness of the sheet.

Moreover, for G > 100, Equation 4.13 shows that there is a vertical tangent at the origin of the curve $\frac{M_t}{M_\infty}$ vs. time, and that a straight line passing through the origin of time is obtained by plotting $\frac{M_t}{M_\infty}$ vs.$\sqrt{t}$, whose slope is proportional to the diffusivity.

4.1.2.2 Application to Experiments and Calculation

The diffusion parameters, such as the diffusivity D and the coefficient of convective transfer h, have been calculated for the three experimental curves drawn in Figure 4.1 obtained with the following drugs: theophylline, diltiazem HCl, and caffeine, each dispersed through a polymer matrix (ethyl cellulose Ethocel standard 10 premium) with various percent plasticizer (ATBC: acetyl tributyl citrate Citroflex A-4 or DBS: dibutyl sebacate).

As shown in Figure 4.1, the good agreement between the experimental data and the theoretical curves when Equation 4.7 is used for calculation proves the validity of the diffusion model for this drug release (correlation coefficient $r^2 = 0.999$).

The values of the parameters of diffusivity are collected in Table 4.1 [11].

These films were prepared by solvent casting. After all the components were dissolved in 2-propanol and the solution was poured into Teflon molds, the solvent was evaporated slowly (three days at room temperature, followed by three days at 35°C and four days at 50°C). In vitro experiments were carried out using the USP XXIII rotating paddle at 37°C and a stirring rate of 100 rpm in the dissolution medium 0.1 M pH 7.4 phosphate buffer of 250 ml. At intervals, 5 ml of liquid were taken for analysis by UV spectrometry and replaced by 5 ml of fresh medium.

Other results appear in Figure 4.2, which shows the kinetics of drug release from dosage forms made of ethyl cellulose containing various percent plasticizer. From these curves, where the percent plasticizer is varied from 10 to 25%, it clearly appears that in all these cases the process of drug release is controlled by diffusion with a constant diffusivity and that Equation 4.7, based on diffusion–convection, fits well with the experimental data. Secondly, the rate of drug release increases strongly with the plasticizer percentage. In Figure 4.2, the abscissa is obtained by putting $\frac{t}{L^2}$ instead of time, so the only difference between these curves results from a change in the diffusivity.

Finally, master curves are drawn in Figure 4.3 by plotting the amount of drug released at time t as a fraction of the total drug released, as a function of the dimensionless time $\frac{\sqrt{Dt}}{L}$. Various curves are shown with selected values of the other dimensionless number $G = \frac{hL}{D}$.

Of course, for higher values of G (e.g., greater than or equal to 100), a straight line passing through the origin of time is obtained, and Equation 4.10 and Equation 4.11 can be used for calculation. With a very low value of G (e.g., lower than 5), the process of drug release is essentially controlled by convection at the dosage form–liquid interface, whereas for values of G ranging from 5 to 100, the process is controlled by both diffusion and convection, and Equation 4.7 stands for expressing the kinetics.

By plotting the data (e.g., $\frac{M_t}{M_{in}}$ as a function of the square root of time) for a given thickness of the sheet, the value of the diffusivity is obtained from the slope of the curve. Moreover, the tendency for the straight line not to pass through the origin at low times gives an idea of the value of the coefficient of convective transfer

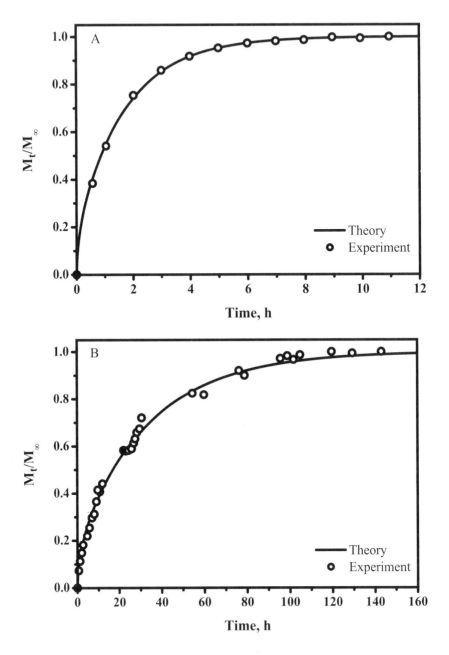

FIGURE 4.1 Experimental (dotted line) and theoretical (solid line) kinetics of drug release with dosage forms made of films. (Fraction of the initial drug released vs. time). Because of the large differences among the times of dissolution, these curves are drawn in three parts. A: Theophylline in a 55 µm thick film made of EC/ATBC (20%w/w). B: Diltiazem HCl in a 173 µm thick film made of EC/DBS (10%w/w). C: Caffeine in a 331 µm thick film made of EC/ATBC (20%w/w). Reprinted from J. Pharmac. Sci. 87(7), pp. 827–832, 1998. With permission From Wiley-Liss, Inc., a Wiley company.

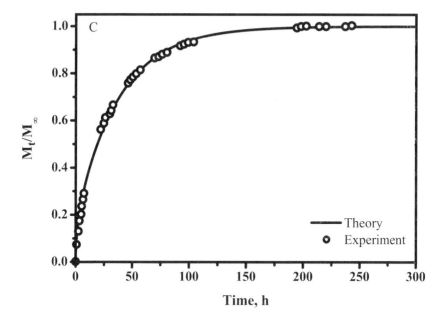

FIGURE 4.1 (Continued)

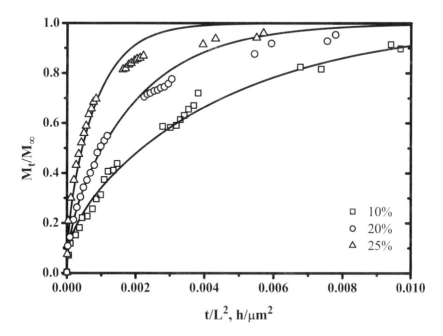

FIGURE 4.2 Experimental (dotted line) and theoretical (solid line) kinetics of drug release with dosage forms made of EC films (diltiazem HCl in EC films) with various percentages of plasticizer (DBS): 25%, 20%, 10% (the values are shown in the figure). Reprinted From J. Pharmac. Sci. 87(7), pp. 827–832, 1998. With permission From Wiley-Liss, Inc., a Wiley company.

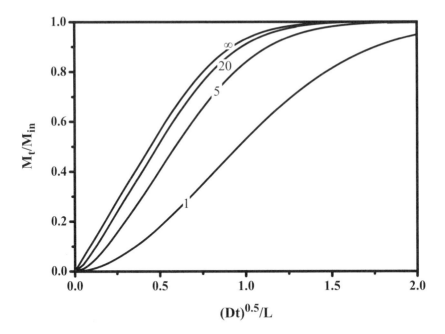

FIGURE 4.3 Master curves expressing the kinetics of release of a drug from a film of thickness 2L, controlled by diffusion, with various values of the coefficient of convection (the values of the dimensionless number $G = \frac{hL}{D}$ are noticed, ranging from 1 to ∞).

at the dosage form–liquid interface. The interest in determining the diffusivity of a drug–polymer system is obvious: As the time required is proportional to the square of the thickness, brief experiments can be carried out by using a sufficiently low thickness.

4.1.2.3 Consideration on the Drug–Polymer Couple

Because the system described previously consists of three various components (such as the drug, polymer, and plasticizer), any basic examination of the phenomenon is highly complex [11, 12], and various approaches could be considered. Two facts clearly appear:

- A larger diffusivity for caffeine than for theophylline
- An increase in the diffusivity with the percent plasticizer

Explanation using the free-volume theory (based on the assumption that the dimension of the diffusing substance plays the main role) fails, because the caffeine molecule is larger than the theophylline molecule. In fact, in this case, the volume of the molecule of the liquid entering the polymer, rather than that of the drug, would be considered [5, 6]. It is also assumed that the drug is in molecular state well dispersed through the polymer [11], but this assumption may be uncommon if we consider the process of drying [13]: When the solvent evaporates, the salt or drug crystallizes from the solution, and the size of the crystallites depends on the rate of drying—the slower the rate of drying, the bigger the crystallites. So is the presence of the polymer of any concern in the crystallization of the drug or of the polymer itself? Now, how effective is the concept of interactions among all these components, and especially the solubility of the drug in the polymer or even in the plasticizer? The rate of release of the drug increases with the percent plasticizer. Thus, must we think that a salt like diltiazem HCl is more soluble in a lipophilic plasticizer than in a hydrophilic polymer such as ethyl cellulose, whereas the synthetic intestinal liquid may be poorly soluble in the plasticizer?

These basic questions should not make us forget the main result: The process of diffusion with a film leads to the basic conclusion that the rate of drug release is proportional to the square of the thickness of the sheet.

4.1.3 EFFECT OF THE SHAPE OF DOSAGE FORMS OF SIMILAR VOLUME

It is of interest to know the precise effect of the shape given to dosage forms of similar volume on the kinetics of release of the drug obtained under similar stirring conditions in the same synthetic gastrointestinal liquid.

The following shapes are considered in succession (Figure 4.4):

- A parallelepiped
- A cube
- A cylinder
- A sphere with a radius of 0.182 cm

4.1.3.1 Kinetics of Release from the Sphere

For a radial diffusion and a constant diffusivity, the diffusion equation takes the following form:

$$\frac{\partial C}{\partial t} = \frac{D}{r^2} \frac{\partial}{\partial r} \left[r^2 \frac{\partial C}{\partial r} \right] \tag{4.16}$$

Upon putting in the function U

$$U = Cr \tag{4.17}$$

the equation for U becomes

$$\frac{\partial U}{\partial t} = D \frac{\partial^2 U}{\partial r^2} \tag{4.18}$$

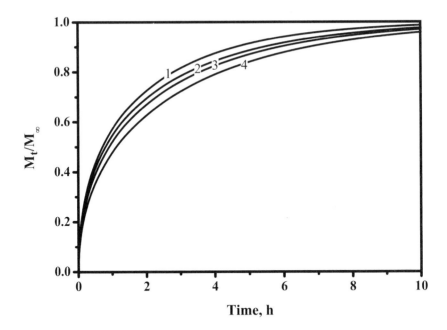

FIGURE 4.4 Curves expressing the kinetics of release of a drug from dosage forms of same volume and different shapes, controlled by diffusion. 1. Parallelepiped with $2a = 2b = c$. 2. Cube of sides $2a$. 3. Cylinder with $R = H$. 4. Sphere with $R = 0.182$ cm. $D = 2.5 \times 10^{-7}$ cm^2/s.

with the following initial and boundary conditions:

$$t = 0 \qquad r < R \qquad U = rC_{in} \tag{4.19}$$
$$t > 0 \qquad r = 0 \qquad U = 0$$

$$t > 0 \qquad r = R \qquad U = RC_{\infty} \tag{4.20}$$

The equation for U corresponds to the problem of diffusion through a plane sheet of thickness R with constant concentrations on each surface (e.g., at $r = 0$, $U = 0$; at $r = R$, $U = RC_{\infty}$), while the initial distribution is rC_{in}. The solution to this problem is given in the case of the membrane [4]. The amount of drug released after time t, M_t, as a fraction of the corresponding quantity after infinite time, M_{∞}, is expressed in terms of time by this equation:

$$\frac{M_{\infty} - M_t}{M_{\infty}} = \frac{6}{\pi^2} \sum_{n=1}^{\infty} \frac{1}{n^2} \exp\left[-\frac{n^2 \pi^2}{R^2} Dt \right] \tag{4.21}$$

When the liquid is not stirred, there is a surface condition:

$$-D \frac{\partial C}{\partial r} = h(C_s - C_{ext}) \tag{4.22}$$

and the kinetics of drug release is shown as follows:

$$\frac{M_\infty - M_t}{M_\infty} = \sum_{n=1}^{\infty} \frac{6K^2}{\gamma_n(\gamma_n^2 + K^2 - K)} \exp\left[-\frac{\gamma_n^2}{R^2} Dt\right] \tag{4.23}$$

with the dimensionless number K in a form similar to G:

$$K = \frac{hR}{D} \tag{4.24}$$

where the γ_ns are the roots of

$$\gamma_n \cot \gamma_n = 1 - K \tag{4.25}$$

Some roots of Equation 4.25 are given for various values of K [1–4], but the best approach is to use a program, because it is impossible to give these values for every K.

4.1.3.2 Calculation for the Parallelepiped of Sides 2a, 2b, 2c

The amount of drug released from the parallelepiped after time t, M_t, as a fraction of the corresponding amount after infinite time, M_∞, is expressed in terms of time by the product of the three series corresponding with one-dimensional transport through the three sheets of thickness $2a$, $2b$, $2c$:

$$\frac{M_\infty - M_t}{M_\infty} = \frac{512}{\pi^6} \sum_{m=0}^{\infty} \frac{1}{(2m+1)^2} \exp\left[-\frac{(2m+1)^2 \pi^2}{4a^2} Dt\right] \sum_{n=0}^{\infty} \frac{1}{(2n+1)^2}$$

$$\times \exp\left[-\frac{(2n+1)^2 \pi^2}{4b^2} Dt\right] \sum_{p=0}^{\infty} \frac{1}{(2p+1)^2} \exp\left[-\frac{(2p+1)^2 \pi^2}{4c^2} Dt\right] \tag{4.26}$$

For brief times of release, when $\frac{M_t}{M_\infty} < 0.6$, this simple relation is obtained:

$$\frac{M_t}{M_\infty} = \left(\frac{2}{a} + \frac{2}{b} + \frac{2}{c}\right)\sqrt{\frac{Dt}{\pi}} - \left(\frac{4}{ab} + \frac{4}{ac} + \frac{4}{bc}\right)\frac{Dt}{\pi} + \frac{8}{abc}\left(\frac{Dt}{\pi}\right)^{1.5} \tag{4.27}$$

For long times of release, when $\frac{M_t}{M_\infty} > 0.7$, the first term in the exponential series becomes preponderant, and the amount of drug released can be expressed in terms of time by

$$\frac{M_\infty - M_t}{M_\infty} = \frac{512}{\pi^6} \exp\left[-\left(\frac{1}{a^2} + \frac{1}{b^2} + \frac{1}{c^2}\right)\frac{\pi^2}{4} Dt\right] \tag{4.28}$$

4.1.3.3 Calculation for the Cube of Side 2a

For the cube of sides $2a$, Equation 4.26, with $2a = 2b = 2c$, reduces to

$$\frac{M_\infty - M_t}{M_\infty} = \frac{512}{\pi^6} \left[\sum_{m=0}^{\infty} \frac{1}{(2m+1)^2} \exp\left(-\frac{(2m+1)^2 \pi^2}{4.a^2} Dt\right)\right]^3 \tag{4.29}$$

For the cube of sides $2a$, the kinetics of drug release for brief times is easily obtained by putting $2a = 2b = 2c$ into Equation 4.27:

$$\frac{M_t}{M_\infty} = \frac{6}{a}\sqrt{\frac{Dt}{\pi}} - \frac{12}{a^2}\frac{Dt}{\pi} + \frac{8}{a^3}\left(\frac{Dt}{\pi}\right)^{1.5} \tag{4.30}$$

For long times, Equation 4.28 or Equation 4.29 reduces to

$$\frac{M_\infty - M_t}{M_\infty} = \frac{512}{\pi^6} \left[\exp\left(-\frac{\pi^2}{4a^2} Dt\right)\right]^3 \tag{4.31}$$

4.1.3.4 Calculation for a Cylinder of Radius R and Height 2 H

In a cylinder of finite length, diffusion is both radial and longitudinal. The solution to the problem is thus expressed by the product of the solutions obtained either for radial diffusion only or for longitudinal diffusion only.

For the infinite coefficient of convection at the surface, the solution is

$$\frac{M_\infty - M_t}{M_\infty} = \frac{32}{\pi^2}\sum_{n=1}^{\infty}\frac{1}{q_n^2}\exp\left(-\frac{q_n^2}{R^2} Dt\right)\sum_{p=0}^{\infty}\frac{1}{(2p+1)^2}\exp\left(-\frac{(2p+1)^2\pi^2}{4H^2} Dt\right) \tag{4.32}$$

where the q_n s are the roots of the Bessel function of the first kind of order 0:

$$J_0(q_n R) = 0 \tag{4.33}$$

These roots of the Bessel function are given in the tables [1–4].

The kinetics of drug release from the dosage forms of same volume and different shapes controlled by diffusion drawn in Figure 4.4 lead to a few comments of interest:

- The kinetics of drug release depends slightly on the shape given to the dosage forms; it is faster for the parallelepiped, resulting from the presence in this solid of the thinner thicknesses of two of its three sides, and it is slower for the sphere.
- The shapes of these kinetics are typical; starting with a high rate at the beginning of the process associated with a nearly vertical tangent, the rate of drug release decreases with time in an exponential way.
- Theoretically speaking, a drawback appears with the fact that the whole drug initially in dosage form is released after infinite time, resulting from the diffusion process.

4.1.4 KINETICS OF DRUG RELEASE FROM SPHERICAL DOSAGE FORMS

Spherical beads are commonly used, either in the form of a solid or in small granules gathered in an envelope. The kinetics of drug release can be determined through in vitro experiments by using either Equation 4.21 (when the synthetic liquid is strongly stirred) or Equation 4.22 through Equation 4.25 (when the coefficient of convective transfer at the interface must be taken into account). In the latter case, it is better to use a typical program.

When the data concerning the diffusion are already known (e.g., the diffusivity D and the coefficient of convective transfer h), the curves expressing the amount of drug released as a function of time can be obtained by calculation. This is the case shown in Figure 4.5, where the profiles of theophylline released are drawn as a function of time for various values of the radius of the microparticles [11]. Of course, a faster release appears with smaller particles, as the dimensionless number $\frac{Dt}{R^2}$ holds. Because the diffusivity is the same for these three types of sphere, the time necessary for a given amount of drug released is proportional to the square of the radius. These microparticles were prepared by using ethyl cellulose and 20% ATBC as plasticizer, and the parameters of diffusion determined using thin sheets made of the same material and shown in Table 4.1 were used for calculation. In fact, the coefficient of convective transfer at the sphere–liquid interface was considered infinite, allowing Equation 4.21 to be used.

The effect of the parameters of diffusion (e.g., the diffusivity D and the coefficient of convection h), as well as the radius of the sphere, is shown in Figure 4.6, where the kinetics of drug release is drawn by using the following dimensionless number: the amount of drug released at time t as a fraction of the corresponding amount after infinite time (which, in fact, is the amount of drug initially located in the dosage form) as ordinate, and the dimensionless time $\frac{\sqrt{Dt}}{R}$ as abscissa. The various kinetics depicted for various values of the dimensionless number $K = \frac{hR}{D}$ clearly show the effect of the coefficient of convective transport on the bead surface. For K larger than 100 or even 50, the process is controlled by diffusion and Equation 4.21 can be used; when K is lower than 5, the process is controlled by convection; when K is between 5 and 50, the process is controlled by both diffusion and convection, necessitating

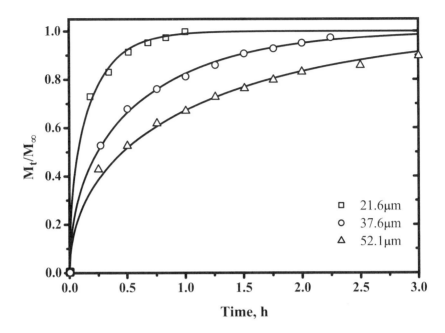

FIGURE 4.5 Experimental (dotted lines) and theoretical (solid lines) kinetics of drug release from dosage forms made of microspheres, controlled by diffusion. Theophylline in EC/ATBC (20% w/w). Effect of the value of the radius (μm). 1. 21.6. 2. 37.6. 3. 52.1. Reprinted From J. Pharmac. Sci. 87(7), pp. 827–832, 1998. With permission From Wiley-Liss, Inc., a Wiley company.

the use of the group of Equation 4.22 through Equation 4.25. In all cases, the amount of drug released after any time is proportional to the square of the radius of the spherical dosage form.

4.1.5 KINETICS OF RELEASE WITH PARALLELEPIPED AND CUBIC FORMS

The equations for the kinetics of release have been established and written for the parallelepiped form obtained with strong stirring, either along the whole process (Equation 4.26) or with brief times (Equation 4.27) and long times (Equation 4.28). The diffusion being longitudinal along the three perpendicular axes, the general solution is expressed in terms of the product of the series obtained for the one longitudinal diffusion shown in a sheet.

When the stirring is not strong, the problem is slightly more complex because the equations are very big, but the same process holds: The solution is the product of each series obtained for a sheet, as shown in Equation 4.7 through Equation 4.9:

$$\frac{M_\infty - M_t}{M_\infty} = f(D,a,h)f(D,b,h)f(D,c,h) \tag{4.34}$$

where all the three functions of the form $f(D,a,h)$ are described in Equation 4.7.

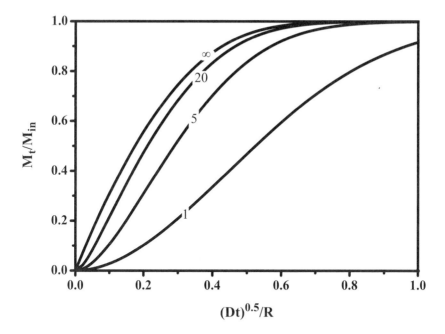

FIGURE 4.6 Master curves for the kinetics of release from spheres of radius R, controlled by diffusion, with various values of the coefficient of convection and $K = \frac{hR}{D}$ (various values of K are noticed).

In Figure 4.7, the amount of drug transferred after time t as a fraction of the corresponding amount after infinite time (in fact, the initial amount in the dosage form) is expressed in terms of the dimensionless time for parallelepipeds of various lengths. In each case, the two sides a and b are constantly equal; only side c is varied from a to $4a$.

The following conclusions deserve mention:

- Of course, the volume of the dosage form increases proportionally with the length given to side c.
- An increase in the length of side c is responsible for a slight increase in time of release. This slight change in time results from the fact that the small sides $2a$ and $2b$ participate more strongly in the kinetics of release than the longer side $2c$ can.
- This result contrasts with the results obtained with a parallelepiped and a cube of same volume (Section 4.1.2 and shown in Figure 4.4), where the kinetics of release was faster for the parallelepiped. This fact results from the thinner sides of the parallelepiped associated with the increase in the length necessary to keep the volume of the parallelepiped constant.
- Of course, when the three sides are equal, the cube is obtained.

For a cubic dosage form, the mathematical treatment is simpler. By recalling that the kinetics of release from a cube is equal to the cube of the kinetics obtained for a sheet of the same side a, Equation 4.34, obtained for the parallelepiped,

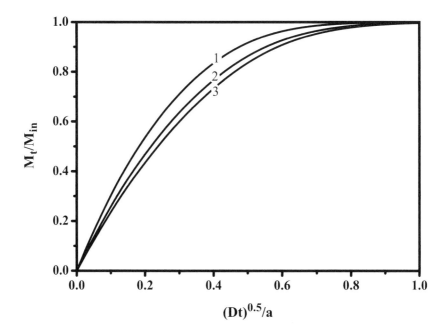

FIGURE 4.7   Master curves expressing the kinetics of release of a drug from parallelepipeds of various lengths, controlled by diffusion.1. $a = b = c$. 2. $c = 2a = 2b$. 3. $c = 4a = 4b$.

reduces to:

$$\frac{M_\infty - M_t}{M_\infty} = f^3(D,a,h) \tag{4.35}$$

where the function f is given for a sheet either in Equation 4.7 (when the rate of stirring is low) or in Equation 4.10 (when the rate of stirring is very strong).

From the curves drawn in Figure 4.8, the conclusions are similar to those made for the sheet:

- For high values of the coefficient of convection h, leading to G values larger than 50, the process is controlled by diffusion.
- For values of G between 5 and 50, the process is controlled by both diffusion and convection.
- A strong increase in the rate of release is obtained in reducing the sides of the cube, because the dimensionless number $\frac{Dt}{a^2}$ intervenes three times. However, when the thickness of a sheet is reduced to half its value, there is a fourfold increase in the rate in response; for a cube, when the sides are reduced to their halves, there is a twelvefold increase in the rate, as proved in Equation 4.28 or more clearly in Equation 4.30. (In Equation 4.30, the time is proportional to $\frac{a}{6}$ for the cube instead of $\frac{a}{2}$ for the sheet in Equation 4.13.)

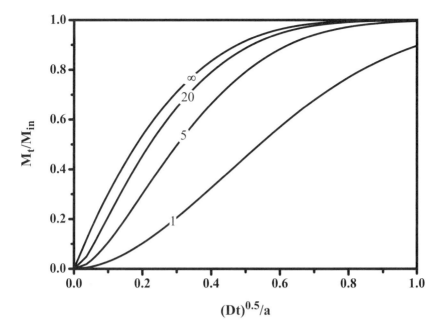

FIGURE 4.8 Master curves expressing the kinetics of release of a drug from cubic dosage forms, controlled by diffusion, with various values of the coefficient of convection and of the dimensionless number $\frac{ha}{D}$ (these values are noticed).

4.1.6 CYLINDRICAL DOSAGE FORMS

For cylindrical dosage forms, two dimensions intervene with the radius R and the length $2H$, not forgetting the parameters of diffusion (such as the diffusivity and the coefficient of convection).

As already shown (Section 4.1.2) with Equation 4.32 and Equation 4.33, the solution for the kinetics of drug release is the product of two series: one expressing the effect of the radial diffusion and the other that of the longitudinal diffusion.

The effect of length, noted $2H$, is especially considered in Figure 4.9, which shows the kinetics of drug release for cylinders of various lengths. These kinetics are calculated by using Equation 4.32 and Equation 4.33 with a high coefficient of convection. Dimensionless numbers are used, either for the ordinate, with the amount of drug released after time t as a fraction of the corresponding amount after infinite time (or the drug initially located in the dosage form), or for the abscissa with $\frac{\sqrt{Dt}}{R}$ because the radius remains the same. A couple of remarks can be made:

- Obviously, the amount of drug released and the amount of drug initially in the dosage form are proportional to the length when the radius is kept constant.
- By using the dimensionless number $\frac{M_L}{M_\infty}$, the kinetics of drug release depicted in Figure 4.9 shows the slight effect of the length when it is increased by two or four.

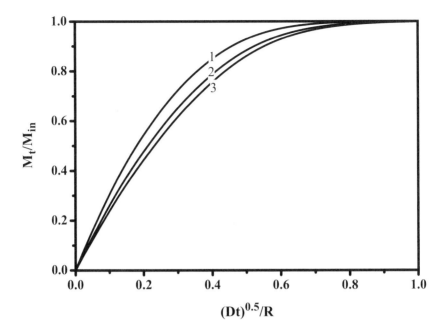

FIGURE 4.9 Master curves expressing the kinetics of release of a drug from dosage forms, cylindrical in shape, controlled by diffusion, with various lengths. 1. $R = H$. 2. $H = 2R$. 3. $H = 4R$.

The effect of the rate of stirring is shown in Figure 4.10, where the kinetics of drug release is drawn with various values of the dimensionless number $\frac{hR}{D}$, and the other two dimensionless numbers are used in the coordinates. These curves lead to the conclusion that, for a cylinder having a height equal to its diameter, the effect of the rate of stirring clearly appears. When the dimensionless number $\frac{hR}{D}$ is very high (e.g., larger than 50), a straight line is obtained; when this ratio is lower than 5, the process is controlled by convection; and when this ratio is between 5 and 50, an S-shape curve can be observed.

4.1.7 DETERMINATION OF THE PARAMETERS OF DIFFUSION

If the diffusivity is rather easy to determine, especially in the case of a sheet—or even a parallelepiped or a sphere—when the rate of stirring is high, the evaluation of the parameters of diffusion (e.g., the diffusivity and the coefficient of convection) becomes a more difficult problem when the stirring rate is low.

By plotting the amount of drug released after time t as a fraction of the amount of drug initially in the dosage form, in terms of the square root of time, a straight line is obtained when the rate of stirring is high, and the diffusivity can be calculated from the slope. For instance, for a sheet of thickness $2L$, according to

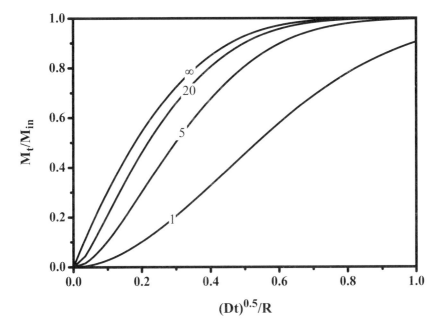

FIGURE 4.10 Master curves expressing the kinetics of release of a drug from dosage forms, cylindrical in shape with $R = H$, controlled by diffusion, with various values of the coefficient of convection and of the dimensionless number $\frac{hR}{D}$ (the values are noticed).

Equation 4.13, the slope of the line is expressed in terms of the diffusivity as follows:

$$\text{slope} = \frac{2}{L}\sqrt{\frac{D}{\pi}} \tag{4.36}$$

When the rate of stirring is low, which is responsible for an S-shape for the kinetics of drug release and for the preceding line not to pass through the origin of time, it is better to use a mathematical program.

Another simple approach consists of evaluating the diffusivity from Equation 4.14, which can be applied to a sheet of thickness $2L$ but also to a sphere of radius R:

$$\frac{Dt_{0.5}}{L^2} = 10\frac{Dt_{0.5}}{R^2} = 0.196 \tag{4.37}$$

where $t_{0.5}$ is the time necessary for the release of half the amount of drug initially located in the dosage form.

4.2 DRUG RELEASE CONTROLLED BY EROSION

In this chapter, dosage forms of various shapes are made of the same drug dispersed in the same erodible polymer, in order to make easy comparisons.

The rate of erosion remains constant during the process, and it is expressed in the following linear form:

$$v = \frac{dL}{dt} \tag{4.38}$$

The initial concentration of the drug is uniform in the dosage form. The kinetics of release is determined in the following sections for various shapes given to the dosage forms.

4.2.1 KINETICS OF RELEASE FOR A SHEET OF THICKNESS $2L_0$

The sheet of thickness $2L_0$ at time 0 is immersed on both sides in the liquid, and at time t, half the thickness becomes L_t, given by

$$L_t = L_0 - vt \tag{4.39}$$

It is completely eroded after the time of erosion t_r:

$$L_t = 0 \quad \text{for} \quad t_r = \frac{L_0}{v} \tag{4.40}$$

By introducing the time of erosion, Equation 4.39 becomes

$$L_t = L_0\left[1 - \frac{t}{t_r}\right] \tag{4.41}$$

And finally, the amount of drug released at time t, M_t, is proportional to time:

$$\frac{M_t}{M_{in}} = \frac{t}{t_r} \tag{4.42}$$

where M_{in} is the amount of drug initially in the sheet.

4.2.2 KINETICS OF RELEASE FOR A SPHERE OF INITIAL RADIUS R_0

At time t, the new radius is given by

$$R_t = R_0 - vt = R_0\left[1 - \frac{t}{t_r}\right] \tag{4.43}$$

The amount of drug released up to time t, M_t, is expressed as a fraction of the amount initially in the sphere, in terms of time:

$$\frac{M_t}{M_{in}} = 1 - \left[1 - \frac{t}{t_r}\right]^3$$

(4.44)

4.2.3 KINETICS OF RELEASE FOR A CYLINDER OF RADIUS R_0 AND HEIGHT $2H_0$

The amount of drug released after time t, M_t, is

$$\frac{M_t}{M_{in}} = 1 - \left[1 - \frac{vt}{R}\right]^2 \left[1 - \frac{vt}{H}\right]$$

(4.45)

When $H < R$, the time of full erosion is given by

$$t_r = \frac{H}{v}$$

(4.46)

and the amount of drug released after time t is

$$\frac{M_t}{M_{in}} = 1 - \left[1 - \frac{H}{R}\frac{t}{t_r}\right]^2 \left[1 - \frac{t}{t_r}\right]$$

(4.47)

4.2.4 KINETICS OF RELEASE FOR A PARALLELEPIPED OF SIDES $2A$, $2B$, AND $2C$

For this parallelepiped, the amount of drug released after time t, M_t, is given by

$$\frac{M_t}{M_{in}} = 1 - \frac{(a - vt)(b - vt)(c - vt)}{abc}$$

(4.48)

When the sides are such that $a < b < c$, the time of full erosion is given by

$$t_r = \frac{a}{v}$$

(4.49)

and the amount of drug released after time t is

$$\frac{M_t}{M_{in}} = 1 - \left[1 - \frac{t}{t_r}\right]\left[1 - \frac{a}{b}\frac{t}{t_r}\right]\left[1 - \frac{a}{c}\frac{t}{t_r}\right]$$

(4.50)

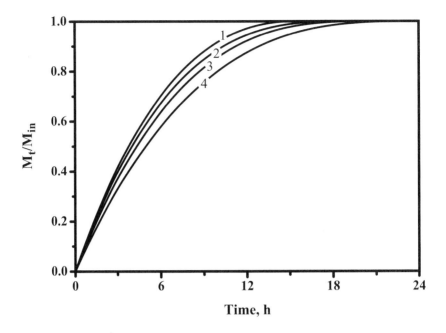

FIGURE 4.11 Kinetics of release of a drug from dosage forms of same volume and different shapes, controlled by erosion. 1. Parallelepiped with $2a = 2b = c$. 2. Cube. 3. Cylinder with $R = H$. 4. Sphere. Time of full erosion = 24 h for the sphere.

4.2.5 KINETICS OF RELEASE FOR A CUBE OF SIDES 2A

For the cube of sides $2a$, the amount of drug released after time t is obtained from Equation 4.50 when the sides are equal to $2a$:

$$\frac{M_t}{M_{in}} = 1 - \left[1 - \frac{t}{t_r} \right]^3 \tag{4.51}$$

Equation 4.51 (for the cube) and Equation 4.44 (for the sphere) are similar at first glance, but the time of erosion is not the same, being proportional to either the radius of the sphere or half the edge of the cube. It can be said that half the edge of the cube, a, is not more than 0.806 times the value of the radius of the sphere having the same volume, meaning that the time of erosion is 0.806 times shorter for the cube than for the sphere. This is why the kinetics of erosion is faster for the cube than for the sphere, as shown in Figure 4.11. In the same way, the time of erosion is proportional to the thinner dimension of a solid, and this fact explains that the kinetics of erosion is faster for the parallelepiped than for a cube of same volume.

4.3 COMPARISON AND OTHER KINETICS OF RELEASE

4.3.1 COMPARISON OF THE PROCESSES OF DIFFUSION AND EROSION

Comparison of the results obtained with the kinetics of drug release controlled either by diffusion or by erosion is of great concern. The curves expressing the kinetics of drug release are shown in Figure 4.4 (when they are controlled by diffusion) or in Figure 4.11 (when they are controlled by erosion). The following conclusions are worth noting:

- The slopes of these curves are quite different. The rate of drug release controlled by diffusion is very high at the beginning of the process, and it decreases exponentially with time. On the other hand, the rate of drug release controlled by erosion starts rather slowly and decreases slowly with time.
- More precisely, the process controlled by diffusion is associated with kinetics that is far from being linear with time, whereas the erosion process leads to kinetics nearly linear with time. Thus, for a sheet it can be said that a straight line would be obtained with erosion.
- With the process of erosion, there is a time of full erosion, well defined, meaning that all the drug initially in the dosage form is totally released up to that time. In contrast, the process of drug release controlled by diffusion comes to an end after infinite time, at least mathematically speaking.
- Not only the drug but also the excipient and additives are released and dissolved after a finite time, so dosage forms whose release is controlled by erosion can be used as adhesive forms either on the stomach or the intestinal wall. As proved in Chapter 6 and Chapter 7, great interest could be found in getting a very long release time for the dosage form (e.g., several days) with a nearly constant rate of drug delivery. A few polymers are capable of swelling and binding with the macromolecules of the mucus. Some of these are erodible, and we may notice in the family of the polysaccharides, the alginate, carraghenate, dextran, chitosan and starch, and protein derivatives such as gelatin.

4.3.2 PROCESS OF DRUG RELEASE BY DIFFUSION–EROSION

It seems logical that the process of erosion in a liquid does not happen mechanically. The liquid diffuses into the polymer and dissolves both the drug at disposal and the polymer when the local concentration of the liquid in the polymer is high enough. Thus, there is a moving boundary on the surface of the dosage form.

4.3.3 PROCESSS WITH DIFFUSION AND SWELLING

As already proved [5–7], the process of diffusion is not simple, because a drug in solid state does not leave the dosage form whatever the excipient may be. The whole process consists of two matter transfers controlled by diffusion:

- The liquid enters the polymer and dissolves the drug.
- This in turn enables the drug to diffuse through the liquid that has previously entered the polymer.

Thus, the process is still more complex, because the diffusivity for drug transport was found to vary with the concentration of the liquid, in the same way as the drug's solubility.

When the rate of diffusion of the liquid into the dosage form is much larger than the rate of diffusion of the drug out of the dosage form, a swelling takes place. The general process of diffusion with swelling has been considered many times since the late 1980s [14], and researchers hoped, at least for a time, to attain a nearly constant rate of drug release with a simple polymer device and a sufficient knowledge of the process of diffusion and polymer swelling [15]. After a while, however, it was found that this objective could not be reached [16]. In fact, the whole process of diffusion with subsequent swelling is highly complex, as proved in two recent studies. Nevertheless, two dimensionless numbers were introduced [15]:

- **The Deborah number,** which equals the relaxation time of the polymer to the rate of diffusion of the liquid
- **The swelling number,** which compares the motion of the liquid and that of the drug

The interest of these numbers has been pointed out in recent studies [17, 18].

Thus, when the liquid diffuses into the polymer, a swelling occurs, and the change in the boundaries of the solid must be considered, making mathematical treatment infeasible [19]. In addition, the diffusivity becomes concentration dependent, increasing with time or, rather, with the concentration of the liquid, which also increases during the process [20]. Numerical methods should be used and tedious experiments made to resolve the problem. For dosage forms of typical shapes, an anisotropic behavior toward diffusion could also be seen, as was shown with rubber discs immersed in toluene [21, 22].

With excipients made of various polymers, some soluble and others insoluble in the liquid, a much larger swelling of the insoluble polymer takes place when the liquid enters the dosage form and dissolves the drug (as well as other additives), creating large cavities full of liquid through which the drug diffuses very quickly. As a result of this complex phenomenon, the rate of drug release follows neither the process of diffusion nor that of erosion; nevertheless, a nearly constant rate of release is obtained. In this case, the best and simplest approach is to fit a polynomial with the experimental data acquired through in vitro tests, because it is necessary to use a mathematical function for calculating the plasma drug level (Chapter 6 and Chapter 7).

Finally, it is worth noting here the interest of using erodible polymers, which can adhere to the gastrointestinal mucus, so as to prolong the residence time of the dosage form along the GIT, leading to rather constant drug profiles in the blood, as will be considered in Chapter 7. Among the various studies made on these kinds of polymers, we can note a few [23–25].

REFERENCES

1. Crank, J., *The Mathematics of Diffusion,* Clarendon Press, Oxford, 1975, Chapters 1–6.
2. Carslaw, H.S. and Jaeger, J.C., *Conduction of Heat in Solids,* Clarendon Press, Oxford, 1959, Chapters 1–7.
3. Vergnaud, J.M., *Liquid Transport Processes in Polymeric Materials,* Prentice Hall, Englewood Cliffs, NJ, 1991, Chapters 1–3.
4. Vergnaud, J.M., *Controlled Drug Release of Oral Dosage Forms,* Hellis Horwood, Chichester, U.K., 1993, Chapters 1–5.
5. Droin, A. et al., Model of matter transfers between sodium salicylate–Eudragit matrix and gastric liquid, *Int. J. Pharm.,* 27, 233, 1985.
6. Malley, I. et al., Modelling of controlled release in case of Carbopol-sodium salicylate matrix in gastric liquid, *Drug Develop. Ind. Pharm.,* 13, 67, 1987.
7. Messadi, D. and Vergnaud, J.M., Simultaneous diffusion of benzyl alcohol into plasticized PVC and of plasticizer from polymer into liquid, *J. Appl. Polym. Sci.,* 26, 2315, 1981.
8. Siewert, M., Perspectives of in vitro dissolution tests in establishing in vivo/in vitro correlations, *Europ. J. Drug Metabol. Pharmacol.,* 18, 7, 1993.
9. Das, S.K. and Gupta, B.K., Simulation of physiological pH-time profile in in vitro dissolution study. Relationship between dissolution rate and bioavailability of controlled release dosage forms, *Drug Develop. Ind. Pharm.,* 14, 537, 1988.
10. Shah, V.P. et al., Influence of higher rates of agitation on release patterns of immediate-release drug products, *J. Pharm. Sci.,* 81, 500, 1992.
11. Siepmann, J. et al., Calculation of the dimensions of drug–polymer devices based on diffusion parameters, *J. Pharm. Sci.,* 87 (7), 827, 1998.
12. Vergnaud, J.M., Scientific aspects of plasticizer migration from plasticized PVC into liquids, *Polym. Plast. Technol. Eng.,* 20, 1, 1983.
13. Vergnaud, J.M., *Drying of Polymeric and Solid Materials,* Springer-Verlag, London, 1992, Chapters 1–5.
14. Hopfenberg, H.B. and Hsu, K.C., Swelling-controlled constant rate delivery systems, *Polym. Eng. Sci.,* 18, 1186, 1978.
15. Peppas, N.A., Release of bioactive agents from swellable polymers: theory and experiments, in *Recent Advances in Drug Delivery Systems,* eds. Anderson, J.M. and Kim, N.Y., Plenum Publishing Corp., New York, 279, 1984.
16. Peppas, N.A. and Segot-Chicq, Les dispositifs à libération contrôlée pour la délivrance des principes actifs médicamenteux. Modélisation des mécanismes diffusionels, *STP Pharma,* 1, 208, 1985.
17. Brazel, C.S. and Peppas, N.A., Mechanisms of solute and drug transport in relaxing, swellable, hydrophilic glassy polymers, *Polymer,* 40, 3383, 1999.
18. Brazel, C.S. and Peppas, N.A., Modeling of drug release from swellable polymers, *Eur. J. Pharm. Biopharm.,* 49, 47, 2000.
19. Bakhouya-Sabbahi, N., Bouzon, J., and Vergnaud, J.M., Absorption of liquid by a sphere with radial diffusion and finite surface coefficient of matter transfer and subsequent change in dimension, *Polym. Compos.,* 2, 105, 1994.

20. Bakhouya, N., Sabbahi, A., and Vergnaud, J.M., Determination of diffusion parameters for polymer spheres undergoing high volume liquid transfer, *Plastics, Rubber Compos.*, 28 (6), 271, 1999.

21. Azaar, K., Rosca, I.D., and Vergnaud, J.M., Anisotropic behaviour of thin EPDM rubber discs towards absorption of toluene, *Plastics, Rubber Compos.*, 31 (5), 220, 2002.

22. Azaar, K., Rosca, I.D., and Vergnaud, J.M., Anisotropic swelling of thin EPDM rubber discs by absorption of toluene, *Polymer*, 43, 4261, 2002.

23. Rossi, S. et al., Model-based interpretation of creep profiles for the assessment of polymer–mucin interaction, *Pharm. Res.*, 16, 1456, 1999.

24. Riley, R.G. et al., An investigation of mucus/polymer rheological synergism using synthesised and characterised poly(acrylic acids), *Int. J. Pharm.*, 217, 87, 2001.

25. Riley, R.G. et al., An in vitro model for investigating the gastric mucosal retention of ^{14}C -labelled poly (acrylic acid) dispersions, *Int. J. Pharm.*, 236, 87, 2002.

5 Bibliography of In Vitro–In Vivo Correlations

NOMENCLATURE

C Free drug concentration in the blood.
C_{max} Free drug concentration at the peak in the blood.
D Diffusivity.
FDA   Food and Drug Administration.
GI(T)   Gastrointestinal (tract).
x, t Abscissa, time.

5.1 GENERAL SURVEY OF IN VITRO–IN VIVO PROBLEMS

5.1.1 CONCEPT OF SUSTAINED RELEASE

The importance of dosage forms is often undervalued, and this fact has sometimes led to formulations with low therapeutic efficiency. Oral drugs are most suitable for the patient, make outpatient treatment possible, and avoid the additional costs associated with intravenous administration in the hospital. On the other hand, there is increasing evidence that serious infections can adequately be treated with oral drugs, and the wide availability of some of them (e.g., fluoroquinolones) makes them especially interesting for serious indications. These agents, when taken orally, have been found as effective as intravenous therapy in the management of different infective processes [1].

Undoubtedly, the elimination half-life time of the drug is the pharmacokinetics parameter most widely used in clinical practice. New drugs have been developed whose main characteristic is a longer half-life time than those of other drugs with the same therapeutic indications. The main advantage of these potential new drugs is the possibility of allowing a once-daily dosing. This simplification of the therapy facilitates the patient's compliance and, overall, leads to a decrease in costs and a reduction in side effects.

In fact, far more easily than conventional dosage forms, drug delivery systems can be designed to achieve a sustained release of the drug in the blood, leading to a nearly constant plasma drug level. The input of novel and older drugs can be improved by delivering them through typical oral dosage forms with sustained release.

5.1.2 OBJECTIVES OF IN VITRO DISSOLUTION TESTS

In vitro dissolution testing is an important tool for characterizing the biopharmaceutical quality of a dosage form at different stages of the drug development. In the early stages of this drug development, perfect knowledge of in vitro dissolution properties is decisive in selecting from various alternative dosage forms the one that will enable better further development of the drug itself. In vitro dissolution data are helpful in the evaluation and interpretation of possible risks, especially in the case of modified-release dosage forms (e.g., regarding the dose delivery) and the food effects on bioavailability that influence the gastrointestinal conditions. Of course, in vitro dissolution tests are of great importance when assessing changes in the manufacturing process. However, none of these purposes can be fulfilled by in vitro dissolution testing without sufficient knowledge of its in vivo relevance; that is, in vitro–in vivo correlations must be studied. These in vitro–in vivo correlations have been defined in different ways and have been subject to much controversy [2–4]. A meaningful correlation must be quantitative so as to allow interpolation between data, thus making the in vitro model predictive. Quantitative extrapolation is not recommended, due to the limited reliability of most types of correlations on this subject.

Possible reasons for poor in vivo–in vitro associations are the following [2]:

- **Fundamentals:** When in vivo dissolution is not the rate limiting the pharmacokinetics stage, and when no in vitro test can simulate the drug dissolution along the GIT
- **Study design:** With inappropriate in vitro test conditions
- **Dosage form:** When the drug release is not controlled by the dosage form or is strongly affected by the stirring of the synthetic liquid
- **Drug substance:** With a nonlinear pharmacokinetics, for example, a first-pass hepatic effect, an absorption window, a chemical degradation, and a large inter- or intra-individual variability

All these factors are of great concern and should be kept in mind, and especially the intervariability of patients' responses to a drug (Chapter 6).

In order to increase their predictive value, dissolution tests aim to mimic in vivo behavior along the GIT, so the in vitro models are adjusted to simulate physiological conditions as closely as possible [5]. Nevertheless, several examples demonstrate that this procedure can also lead to misinterpretations [1]. Thus, from numerous scientific data, it is concluded that there is no need for close simulation of physiological conditions. However, special care with pH, agitation, ionic strength, and surface tension should give conditions nearly similar to the physiological reality, avoiding artificial data and misinterpretations. High rates of agitation are not necessary and should therefore be avoided, because these can cause a loss of sensitivity in distinguishing real differences in the dissolution rates due to stronger agitation [6, 7].

5.2   IN VITRO–IN VIVO CORRELATIONS
FOR IMMEDIATE-RELEASE DRUGS

The stage of dissolution was originally conceived in 1925 as a way to assure drug absorption through the GI wall, but it took another 25 years to become aware of the great concern of this step. In the mid-1950s [8], the pharmaceutics parameters (such as the plasma peak and duration values of the drug) were found to be related to the intrinsic dissolution rate of the drug. In the following decade, aspirin dissolution rates were correlated with the absorption rate in the plasma [9], and the importance of the agitation rate in the dissolution medium used with in vitro tests was established [7]. Also, a rank order for in vitro–in vivo correlations was obtained on four chloramphenicol preparations [10].

Despite these experiments showing the usefulness of the dissolution test, it was only used as a quality control test during the manufacture of the dosage forms. However, in 1972 the Food and Drug Administration (FDA) was forced to deal with the problem set by the nonbioequivalence of a number of generic digoxin formulations put on the market. This drug, first marketed in the U.S. prior to 1938, was not subject to the drug rules of the 1938 amendments, including the bioavailability regulations proposed by this act. While digoxin was reclassified as a "new drug," its regulation was established in 1972. And for purposes of drug regulation, the dissolution test, based on in vivo–in vitro correlation, has been used as a surrogate for in vivo availability as a condition of marketing. Then, the FDA began a systematic study on drug dissolution and in vivo availability for the purpose of developing other in vitro–in vivo correlations. In 1975, these results led the FDA to awareness that the bioequivalence difficulties were due to poor dissolution of the drug, either in the synthetic liquid or along the GIT [11]. Various marketed products exhibiting significant differences in their dissolution rates were tested in vivo, and in vitro–in vivo correlations were established for drugs delivered orally through immediate-release dosage forms. Finally, the mean in vitro dissolution time was correlated with a single pharmacokinetics parameter, such as AUC or the concentration at the peak C_{max}, which now corresponds to Level C correlation (Section 5.3.1).

By 1977, the principle of poor dissolution in aqueous media within a 60 m time as the primary reason of nonbioequivalence was established. At that time, the FDA proposed the use of in vitro data to predict the bioequivalence of a drug, and the rate of dissolution of the drug should be much larger than its rate of absorption in the plasma [12].

In implementing this policy, the FDA collaborated with the U.S. Pharmacopoeia (USP) to incorporate the dissolution tests into their monographs on solid oral dosage forms with immediate release. In the early 1980s, the FDA felt sufficiently comfortable with this in vitro test run as a surrogate for in vivo bioavailability that it was used for the following:

- Batch-to-batch bioequivalence assurance
- Reformulation from a sugar-to-film coated tablet
- Change in the site of manufacture

The key words in the 1990s were *cost containment* and *harmonization*. In taking action and setting standards, regulatory authorities wanted to make their decision based on the most solid data available: in vitro–in vivo correlations. Of course, drugs with narrow therapeutic indices, complex pharmacokinetics, and interconverting metabolites merit special attention [13].

5.3 IN VITRO–IN VIVO CORRELATIONS WITH SUSTAINED-RELEASE FORMS

5.3.1 ESTABLISHMENT OF LEVELS FOR IN VITRO–IN VIVO CORRELATIONS

The situation for controlled-release dosage forms is quite different from that of immediate release. Thus, it has been said that the database providing a comfort level for immediate-release dosage forms is simply non-existent [13]. Not only are there serious questions concerning the interpretation of the data for controlled-release dissolution, even the significance of the basic test parameters (such as the rate of agitation) is unknown.

In 1984 and in 1988, the FDA managed two workshops to consider the issue, based on their experience acquired with immediate-release dosage forms [14, 15]. But "both workshops concluded that the state of science and technology at that time did not always permit meaningful correlations. Unlike the case of immediate release dosage forms, the correlations, when possible, would have to be product specific." [13]

Because there were no generally recognized acceptable correlations, the FDA required bioavailability testing for the regulatory issues to be imposed on immediate-release dosage forms (e.g., site of manufacturing change, formulation changes, and batch size scale-up). Thus, the FDA imposed a bioavailability requirement as a condition of approval.

In June 1992, the FDA, the Canadian Health Protection Branch, and the USP cosponsored an open hearing in Toronto concerning in vitro and in vivo evaluation of dosage forms, dealing essentially with controlled release, in order to answer the questions of how to establish and how to use in vitro–in vivo correlations for controlled-release dosage forms [16].

Three levels of correlation were defined:

- **Level A:** A unit-to-unit relationship established by comparing the in vitro curve to the input function resulting from deconvolution of the plasma concentration time curve
- **Level B:** Mean in vitro dissolution time compared to either the mean residence time or the mean in vivo dissolution time
- **Level C:** Comparison of a single pharmacokinetic parameter, such as AUC or C_{max}, to mean in vitro dissolution time

From the scientific point of view, Level A correlation is preferred to Level B or Level C (in that order) for the following reasons [13, 17]:

- It uses each plasma level and dissolution time point generated and is therefore reflective of the entire curve.

- Because it is predictive of the in vivo performance of the dosage forms, it is an excellent quality-control procedure.
- The boundaries of the in vitro dissolution curve can be justified on the basis of convolution and deconvolution.

Level A correlation has been reported to describe fairly well the results obtained with an oral osmotic pump [18]. It was also employed to design the Theodur^R controlled-release theophylline formulations [19].

Level B uses statistical moment theory. One problem is that one can get the same result from several different plasma curves.

Level C characterizes single-point correlations and therefore is not reflective of the entire plasma drug concentration time curve.

It has also been said that, because of the growing importance of controlled-release dosage forms, there will be increasing regulatory involvement, from preapproval inspections to supplement review and approval [13]. The biggest obstacle thus far has been a lack of agreement about how to develop and use meaningful correlations. A mapping approach with the following three stages is said to be desirable when Level A cannot be established [12, 19]:

- **Stage 1:** Critical manufacturing variables (coating thickness, composition, compression rate)
- **Stage 2:** Dissolution procedure (type of agitation, pH, nature of the liquid medium)
- **Stage 3:** In vivo response

5.3.2 REQUIREMENTS FOR ESTABLISHING LEVEL A CORRELATIONS

Among the three levels of correlation defined, Level A is of most interest. In this section, Level A will be studied from a mathematical and a biopharmaceutical point of view [17].

Two definitions on in vitro–in vivo correlations have been proposed by a USP working group and by the FDA [17, 20, 21]:

- **USP definition:** The establishment of a relationship between a biological property, or a parameter derived from a biological property produced by a dosage form, and a physiochemical characteristic of the same dosage form.
- **FDA definition:** To show a relationship between two parameters. Typically a relationship is sought between in vitro dissolution rate and in vivo input rate. This initial relationship may be expanded to critical formulation parameters and in vivo input rate.

Several prior conditions must be met before attempting to establish correlations [17]:

- The pharmacokinetic profile of the drug must be linear. If the first-pass hepatic depends on the input rate of the drug in the organism, the drug may have linear pharmacokinetics but nonlinear input kinetics.

- The stage of absorption through the GI wall must be limited by the release of the drug from the dosage form. Level A correlation is useful above all for sustained-release dosage forms in which the biopharmaceutical design is based on the drug release.
- The intra- and intersubject variability of kinetics must be ascertained beforehand, because it is useless to establish a correlation if the variability of the response is very wide.
- Establishment of a Level A correlation is appropriate only after single administration because it is a pharmaceutical parameter. It is necessary that the phenomena observed and the relationships obtained after a single administration persist after the second dose and at the so-called steady state, when the curves are reproducible for each dose.
- It is necessary to administer every dosage form to the same patients, particularly an intravenous injection or a solution, so that convolutions and deconvolutions may be performed.
- Level A correlations must be established as early as possible in the dosage form development.
- Numerous dissolution methods, not limited to those officially prescribed, must be available to ensure proper differentiation and discrimination.

5.3.3 MATHEMATICAL TREATMENT OF IN VITRO–IN VIVO CORRELATIONS

As designated by the definitions proposed by the FDA and USP groups [21], Level A specifies one-to-one correlation. The relationship between in vitro dissolution, expressed by the kinetics $x = f(t)$, and in vivo input kinetics, expressed by the relation $y = f(t)$, would be linear, in the form

$$y = ax + b \qquad (5.1)$$

the intercept b being different from 0 and positive whenever possible.

This term b means that the in vivo input kinetics lags behind the in vitro kinetics; this lag time would correspond to the time necessary for the drug to pass through the GI membrane. Surely this lag time exists, but it would be rather low, not exceeding a few minutes. However, because the process of absorption is usually assumed to be described by a first-order kinetics, this fact means that the time necessary for the drug to go across the GI membrane is zero, or at least very low. In conclusion, Equation 5.1 does not hold well; it stands to reason that at time 0, there is no drug in either the GI or the plasma, whatever the in vitro measurement system. It is thus obvious that b in Equation 5.1 should be 0.

A few examples have been given when the OROS[R] osmotic oral dosage form releases the drug (salbutamol sulfate) [22]. The dissolution kinetics were determined using four different methods:

a: The USP rotating paddle
b: The flow-through cell

c: The index release rate tester

d: The Bio disc

The percentage of drug released in vivo (y) was plotted against the percent drug released in vitro in the four cases. Fitting was attempted for a linear relationship between these two kinetics $y = f(x)$. The correlation measured with the r^2 coefficient, between 0.91 and 0.986, is rather poor, meaning that the linear relation between y and x is not the best one. The shape of curve 1 in Figure 5.1, expressing the in vitro–in vivo correlation with the OROS form obtained by plotting the percent released in vivo vs. the percent released in vitro measured using the USP rotating paddle, retraced from Figure 3a [17], evidently shows that there is no reason to try drawing a straight line when a parabolic curve would better represent the phenomenon.

With the OROS systems, the process of release does not start as soon as the OROS form is put in contact with the liquid; on the contrary, the rate of drug release increases very slowly, and at least 1 h is necessary for the constant rate to be attained, as shown in various studies [22, 23].

Another relation would be worth testing, which considers the fact that the drug should diffuse through the GI membrane. This diffusion process is expressed by the

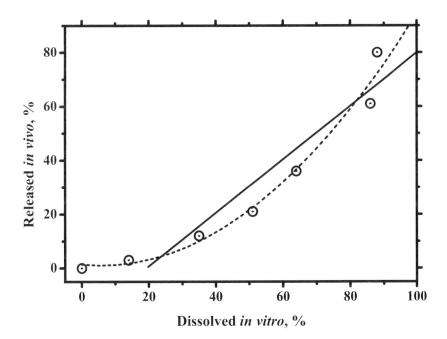

FIGURE 5.1 In vitro–in vivo correlations, with OROS form containing salbutamol sulphate, obtained with the USP rotating paddle. Retraced from Figure 3 [17]. Circles experiments; straight line, from [17]; dotted line, parabolic tendency. Reprinted From Eur. J. Drug metabol. Pharmacok. 18, pp. 113–120, 1993. With permission From Medicine et Hygiene.

following general equation:

$$\frac{\partial C}{\partial t} = D \frac{\partial^2 C}{\partial x^2} \qquad (5.2)$$

with the boundary conditions for the *GI* and the blood compartment:

$$x = 0 \qquad GI \qquad x = f(t) \qquad (5.3)$$

$$x = L \qquad \text{blood} \qquad y = f(t)$$

As shown in Chapter 10, regarding transdermal drug transfer, the kinetics of drug transfer through a membrane is described by Equation 5.2 with the boundary conditions of Equation 5.3. Thus, the amount of drug transferred through the membrane increases slowly with time, as shown in Chapter 10 (Figure 10.3).

Another example is given in Figure 5.2, retraced from Figure 6 in reference [17], with the level A correlation obtained with a suppository form containing indomethacin by plotting the percent drug absorbed vs. the percent dissolved measured using the flow-through cell. It is clear that a straight line cannot express

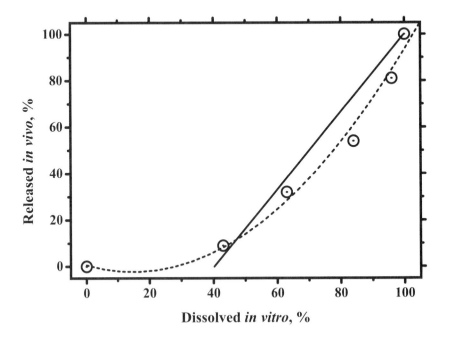

FIGURE 5.2 In vitro–in vivo correlations, with a suppository form containing indomethacin, obtained with the flow-through cell. Retraced from Figure 6 [17]. Circles, experiments; straight line, from [17]; dotted line, parabolic tendency. Reprinted From Eur. J. Drug metabol. Pharmacok. 18, pp. 113–120, 1993. With permission From Medicine et Hygiene.

the phenomenon, as shown with the solid line, but a parabolic curve (dotted line) better represents the correlation.

5.4 CONCLUSIONS

The following concern is important: "It is arguable whether it is justified to find correlations at all costs, by using any in vitro method, even with dissolution media different from physiological conditions, rather than taking a more pragmatic approach towards the feasibility of establishing a Level A correlation" [17].

Other considerations are also of great interest:

- Generally, good correlation is supposed to be obtained when a straight line represents the percent of the drug absorbed vs. the percent dissolved. But it is clear that for either salbutamol sulphate delivered through an OROS system [17,22] or a suppository form containing indomethacin [17,24], the percent released in vitro vs. the percent absorbed in vivo exhibits a kind of parabolic tendency more than a linear relationship, as shown in Figure 5.1 and Figure 5.2, respectively.
- In vitro–in vivo correlations are set up for mean data acquired with healthy volunteers, usually after a single administration, and from averaged data.
- Moreover, although the variability among in vitro results is generally good (5%), the intersubject variability is much larger (more than 100%).

From these two facts, it seems that the safest approach would be to consider this in vitro–in vivo correlation as a development tool for modified-release forms, with possible quality-control applications, rather than an infallible monitoring method [17].

REFERENCES

1. Dominguez-Gil, A., Contribution of biopharmaceutics and pharmacokinetics to improve drug therapy, *Eur. J. Drug Metabol. Pharmacol.*, 18 (1), 1, 1993.
2. Siewert, M., Perspectives of in vitro dissolution tests in establishing in vitro/in vivo correlations, *Eur. J. Drug Metabol. Pharmacol.*, 18 (1), 7, 1993.
3. Hüttenrauch, R. and Speiser, P., In vitro/in vivo correlations: an unrealistic problem, *Pharmacol. Res.*, 3, 97, 1985.
4. Süverkrüp, R., In vitro/in vivo correlations: concepts and misconceptions, *Acta Pharm. Technol.*, 32, 105, 1986.
5. Das, S.K. and Gupta, B.K., Simulation of physiological pH-time profile in in vitro dissolution study. Relationship between dissolution rate and bioavailability of controlled release dosage forms, *Drug Dev. Ind. Pharm.*, 14, 537, 1988.
6. Shah, V.P. et al., Influence of higher rates of agitation on release patterns of immediate-release drug products, *J. Pharm. Sci.*, 81, 500, 1992.
7. Hamlin, W.E. et al., Loss of sensitivity in distinguishing real differences in dissolution rates due to increasing agitation, *J. Pharm. Sci.*, 51, 432, 1962.

8. Nelson, E., Solution rate of theophylline salts and effects from oral administration, *J. Am. Pharm. Assoc. Sci.,* 46, 607, 1957.

9. Levy, G., Comparison of dissolution and absorption rates of different commercial aspirin tablets, *J. Pharm. Sci.,* 52, 1039, 1961.

10. Aquiar, A.J. et al., Evaluation of physical and pharmaceutical factors involved in drug release and availability from chloramphenicol capsules, *J. Pharm. Sci.,* 57, 1844, 1968.

11. Cabana, B.E., In vivo predictability of in vitro dissolution, Annual Industrial Pharmacy Management Conference, Madison, WI, Oct. 27, 1982.

12. Skelly, J.P., Bioavailability and bioequivalence, *J. Clin. Pharmacol.,* 16, 539, 1976.

13. Skelly, J.P. and Shiu, G.F., In vitro/in vivo correlations in biopharmaceutics: scientific and regulatory implications, *Eur. J. Drug Metabol. Pharmacol.,* 18 (1), 121, 1993.

14. Skelly, J.P. et al., Report of the workshop on controlled-release dosage forms: issues and controversies, *Pharm. Res.,* 4, 75, 1987.

15. Skelly, J.P. et al., In vitro and in vivo testing and correlation for oral controlled/modified release dosage forms, *Pharm. Res.,* 7, 975, 1990.

16. Skelly, J.P. et al., Scale-up of immediate oral solid dosage forms, *Pharm. Res.,* 10, 313, 1993.

17. Cardot, J.M. and Beyssac, E., In vitro/in vivo correlations: scientific implications and standardisation, *Eur. J. Drug Metabol. Pharmacol.,*18 (1), 113, 1993.

18. Leeson, L., USP Open Hearing, Toronto, Canada, June 15–18, 1992.

19. Shah, J.P. et al., Scale-up of controlled release products-preliminary considerations, *Pharm. Tech.,* 35, May 1992.

20. USP, Subcommittee on Biopharmaceutics, In vitro–in vivo correlation for extended-release oral dosage forms. Pharmacopoeia Forum, Stimuli to the revision process, USP, 4160, 1988.

21. AAPS/FDA/USP, In vitro and in vivo testing and correlation for oral controlled/modified release dosage forms, Washington, DC, 1988.

22. Civiale, C. et al., In vitro/in vivo correlation of salbutamol sulphate release from a controlled release osmotic pump delivery system, *Methods Find Exp. Clin. Pharmacol.,* 13, 491, 1991.

23. Heilmann, K., *Therapeutic Systems,* 2nd ed., Georg Thieme Verlag, Stuttgart, 1984, 50.

24. Lootvoet, G. et al., Study on the release of indomethacin from suppositories: in vitro–in vivo correlation, *Int. J. Pharm.,* 85, 113, 1992.

6 Plasma Drug Level with Oral Diffusion-Controlled Dosage Forms

NOMENCLATURE

D Diffusivity of the drug through the polymeric matrix of the sustained release dosage form, expressed in cm^2/s.

$\frac{D}{R^2}$ Diffusivity as a fraction of the square of the radius of the sustained release dosage form, spherical in shape, expressed in $second^{-1}$.

GI(T) Gastrointestine or gastrointestinal (tract).

i.v. Intravenous.

H Height of a cylinder.

k_a Rate constant of absorption of the drug (per h).

k_e Rate constant of elimination of the drug (per h).

M_t Amount of drug in the blood at time t.

M_{in} Amount of drug initially in the dosage form.

R Radius of the spherical dosage form.

t Time.

$t_{0.5}$ Half-life time of the drug in bolus i.v. (h).

$T_{0.5}$ Half-life time of the drug obtained in the sustained-release dosage form (h).

V_p Apparent plasmatic volume (l).

$\frac{dM}{dt}$ Rate of drug released by the dosage form in the gastrointestinal volume (Equation 6.1).

W Amount of drug eliminated (Equation 6.3).

Y Amount of drug in the gastrointestinal volume (Equation 6.1).

Z Amount of drug in the plasmatic volume (Equation 6.2).

6.1 METHODS OF CALCULATION

In various papers, the plasma drug level has been calculated with oral dosage forms whose release is controlled by diffusion. The first study was concerned with a spherical dosage form containing theophylline as a drug [1], and the calculation was made by considering the change in the pH by following the pH–time history along the GI transit described by scintigraphy [2]. The second paper, done with a spherical dosage form, dealt with salicylate sodium as the drug dispersed in the polymer Eudragit RL, and the profile of drug concentration was calculated in various parts, such as the GI, the blood volume, the drug eliminated, and the drug released from the dosage form; thus it was found that the amount of drug in the GI was very low [3]. Another study was devoted to the prediction on in vivo blood level with controlled-release dosage forms, by considering especially the GIT time, which is, according to some authors, not more than 10 h [2]; a limitation arises from this fact [4].

The effect of the dose frequency on the plasma drug profile was also regarded [5]. Original dosage forms were also considered as potential issues for delivering the drug in a sustained-release way, by using lipidic Gelucire^R as the excipient [6]. The plasma drug level was also calculated with a more complex system, based on spherical materials made of a core and a shell with a lower concentration in the shell, in order to reduce the drawback of the diffusion process leading to a rate of drug delivery that is very high at the beginning and decreases exponentially with time [7]. Moreover, the results concerning the process controlled by diffusion and by erosion were collected in a general paper and studied more deeply later [8, 9]. The method developed and published in the preceding papers was presented in 1996 to a panel of specialists of the FDA. More recently, the plasma drug level has been related to the characteristics of the dosage forms (such as the dimensions, the nature of the polymer and thus the diffusivity) for various drugs [10].

6.1.1 ASSUMPTIONS FOR CALCULATION

The following assumptions are made in order to define the problem precisely:

- The process of drug transport is divided into the following stages, which are connected to each other (Figure 6.1):
  - Transport through and out of the dosage form into the GI liquid
  - Transport from this liquid to the blood through the GI membrane with the rate constant of absorption
  - Transport from the blood to the surroundings with the rate constant of elimination
- The dosage form, whatever its shape, may or may not keep its dimensions constant during the process, provided that the kinetics of these changes are known.
- The process of drug delivery from the dosage form in the gastric liquid is simplified by considering only the diffusion of the drug through the dosage form, with a constant diffusivity [11]. Nevertheless, this assumption is not mandatory.
- The phenomena of absorption into the blood and elimination from the blood are described by first-order kinetics.
- With in vivo calculation, the rate constant of absorption k_a remains constant along the GIT, as well as the rate constant of elimination k_e, which remains constant during the process. Note that the present numerical model can take into account a dependence of the rate of drug release with the pH and the GIT history (when it is known).

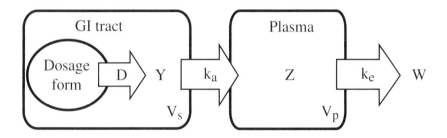

FIGURE 6.1 Scheme of the process of drug delivery with the following stages: release of the drug along the GIT controlled by diffusion, absorption into the plasma compartment, and elimination. The first-pass hepatic is not shown; neither is the binding by the blood protein, which leads to the apparent plasmatic volume.

6.1.2 MATHEMATICAL TREATMENT

The rate of drug delivery by the dosage form at time t, whatever its shape, is given by $\frac{dM}{dt}$. Thus the amount of drug in the GI, Y, is expressed by the relation:

$$\frac{dY}{dt} = \frac{dM}{dt} - k_a Y \qquad (6.1)$$

The amount of drug in the plasma compartment is Z, at time t, and given by

$$\frac{dZ}{dt} = k_a Y - k_e Z \qquad (6.2)$$

The amount of drug eliminated at time t, W, is expressed by

$$\frac{dW}{dt} = k_e Z \qquad (6.3)$$

6.1.3 NUMERICAL TREATMENT

The mathematical treatment of the preceding system of three equations can be resolved only when the drug is released immediately in the GI or when the rate of delivery is constant. In all other cases, a numerical method is necessary to solve the problem. It is run step by step, with a constant increment of time Δt. Two general approaches can be followed:

- Using an old approach with the explicit method
- Using an implicit method, the best-known being the Crank–Nicolson method

The problem is generally complex, and it is better to use a typical program.

By comparing Equation 6.1 through Equation 6.3 with their counterparts Equation 3.1 through Equation 3.3 describing the process with immediate release (Chapter 3), it clearly appears that these equations are similar, except for the first one of each group. In Equation 6.1, the finite rate of drug delivered by the dosage form emerges with $\frac{dM}{dt}$, whereas in Equation 3.1, all the drug is available in the GI to pass through the membrane into the blood compartment. This finite rate $\frac{dM}{dt}$ depends on the dosage form, and all its characteristics play a role: its shape and its dimensions, the nature of the polymeric system that defines the diffusivity, and, secondarily, the nature of the drug when its rate of dissolution is very slow.

6.1.4 Expression of Sustained Release through the Time $T_{0.5}$

For bolus i.v. delivery, the main parameter is the drug's half-life. Similarly, it can be said for oral dosage forms, and especially for sustained-release dosage forms, that the time at which the concentration falls to half its maximum value attained at the peak is also the parameter of interest. This time, noted $T_{0.5}$, should be able to evaluate in quantitatively the effect of the polymeric matrix and the dimensions of the dosage form on the sustained release of the drug.

6.2 RESULTS OBTAINED FOR A SINGLE DOSE

For patients, the body reacts to the drug in a different way, even when the pharmacokinetics is linear, the effect of the metabolism being different and the protein binding leading to the apparent plasmatic volume, which is a typical parameter of each drug. Thus, instead of displaying the plasma drug concentration, which necessitates knowledge of the pharmacokinetic data of the drug, the amount of the drug present in the blood as a fraction of the initial amount of the drug in the dosage form is used in all the figures. Of course, it is easy when the apparent plasmatic volume is known to determine the drug concentration, because the drug amount in the blood is proportional to the concentration.

6.2.1 Assessment of the Amount of Drug in Various Places

The amount of drug, expressed as a fraction of the amount of drug initially in the dosage forms, was calculated in various places and compartments as a function of time, for the following three drugs released from various dosage forms:

- Theophylline in spherical dosage forms (Figure 6.2)
- Caffeine in cylinders (Figure 6.3)
- Diltiazem HCl in parallelepipeds (Figure 6.4)

In order to make comparisons, the curves resulting from immediate release dosage forms are also given. The curves are noted as follows:

- **1, 2, 3:** Drug release kinetics from each of the three dosage forms
- **1', 2', 3':** Drug profiles in the plasma compartment for each drug
- **1", 2", 3":** Kinetics of elimination of each of the three drugs

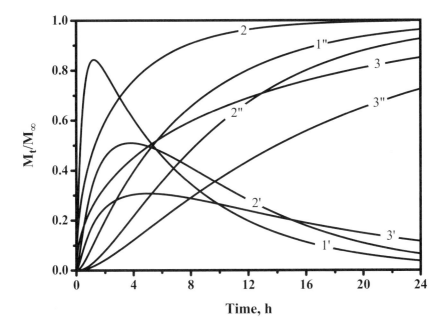

FIGURE 6.2 Relative amount of theophylline vs. time obtained with different dosage forms: immediate release (1, 1', 1"); controlled release from spherical dosage forms with $R = 87$ μm (2, 2', 2"), and with $R = 174$ μm (3, 3', 3"). 1, 2, 3: Kinetics of drug release from the dosage forms. 1', 2', 3': Plasma drug profiles. 1", 2", 3": Kinetics of drug elimination from the plasma. $D = 5 \times 10^{-10}$ cm^2/s; $k_a = 2.5$/h; $k_e = 0.14$/h. Curve 1 superimposes on the y-axis.

The curves 1, 1', and 1" refer to the immediate-release dosage forms, the curves 2, 2', and 2" denote the kinetics obtained with controlled-release dosage forms able to release 85% of their initial drug amount within 6 h, and the curves 3, 3', and 3" represent the kinetics obtained with controlled-release dosage forms able to release 85% of their drug within 24 h.

From the observation of all the curves drawn in Figure 6.2, Figure 6.3, and Figure 6.4, the following conclusions are worth noting:

- The kinetics of drug released from the three dosage forms, expressed by the curves 1, 2, and 3, are quite different. Thus, the curve representing the kinetics of drug release from immediate release is so fast that it is drawn   along the y-axis in the three figures.
- Each drug is dispersed through a different polymeric system, so the diffusivity is different for each drug and written as follows:
  - In Figure 6.2, $D = 5 \times 10^{-10}$ cm^2/s
  - In Figure 6.3, $D = 8.2 \times 10^{-10}$ cm^2/s
  - In Figure 6.4, $D = 2.5 \times 10^{-10}$ cm^2/s
- The dosage forms associated with the curves 2, 2', and 2" are built in such a way that 85% of the drug is released within 6 h, whatever the nature of the polymer and the shape selected: spherical for theophylline (Figure 6.2),

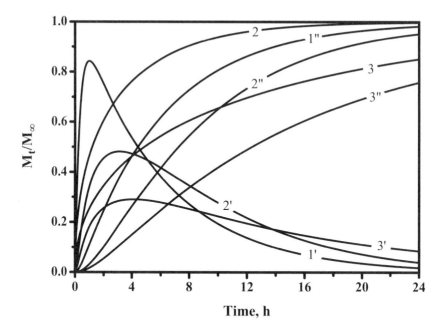

FIGURE 6.3 Relative amount of caffeine vs. time obtained with different dosage forms: immediate release (1, 1', 1"); controlled release from cylindrical dosage forms with $R = 112$ μm and $H = 185$ μm (2, 2', 2"), and with $R = 250$ μm and $H = 320$ μm (3, 3', 3"). 1, 2, 3: Kinetics of drug release from the dosage forms. 1', 2', 3': Plasma drug profiles. 1", 2" 3": Kinetics of drug elimination from the plasma. $D = 8.2 \times 10^{-10}$ cm²/s; $k_a = 3.03$/h; $k_e = 0.17$/h. Curve 1 superimposes on the y-axis. Reprinted from Eur. J. Pharmaceutics Biopharma. 51, 17–24, 2001. With permission from Elsevier.

cylindrical for caffeine (Figure 6.3), and parallelepiped for diltiazem HCl (Figure 6.4). On the other hand, the dosage forms necessary to release 85% of the drug within 24 h (3, 3', 3") are made of the same material and have the same shape as their counterparts built for releasing 85% drug within 6 h but have a larger size. Thus, obviously, the three curves noted 2, expressing the kinetics of release from the dosage forms, are nearly identical in the three figures, and the similarity can be observed in the curves noted 3.

• For each drug, the plasma drug levels shown with the curves 1', 2', and 3' are quite different for each of the three dosage forms. Immediate-release dosage forms are responsible for a high peak attained in a rather short time, whose characteristics can be determined precisely by using Equation 3.8 and Equation 3.9 (Chapter 3). Flat plasma drug profiles are obtained with sustained-release dosage forms, with the obvious statement: the lower the peak, the larger the size of the dosage forms (the time in the process of diffusion being proportional to the square of the main dimension). Moreover, the drug concentration in the plasma associated with these sustained-release dosage forms is nearly constant over a long period of time.

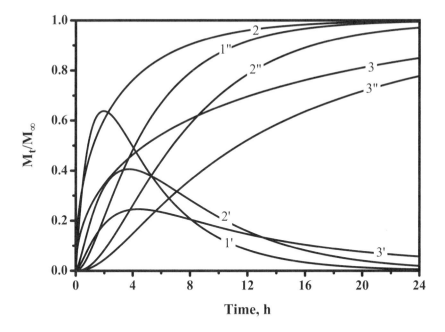

FIGURE 6.4 Relative amount of diltiazem HCl vs. time obtained with different dosage forms: immediate release (1, 1', 1"); controlled release from parallelepiped dosage forms with $96 \times 96 \times 192$ μm (2, 2', 2"), and with $192 \times 192 \times 384$ μm (3, 3', 3"). 1, 2, 3:Kinetics of drug release from the dosage forms. 1', 2', 3': Plasma drug profiles. 1", 2", 3": Kinetics of drug elimination from the plasma. $D = 2.5 \times 10^{-10}$ cm²/s; $k_a = 0.96$/h; $k_e = 0.23$/h. Curve 1 superimposes on the y-axis. Reprinted from Eur. J. Pharmaceutics Biopharma. 51, 17–24, 2001. With permission from Elsevier.

- The kinetics of drug elimination described by the curves 1", 2", and 3" are also typical, depending on the nature of the dosage form. The rate of drug eliminated is lower with the sustained-release dosage forms than with their immediate-release counterparts; this is an advantage when the drug clearance is low.
- The effect of the nature of the drug on the plasma drug level and on the kinetics of elimination can already be appreciated, in spite of the different shapes of the dosage forms selected for each drug. A high value of the rate constant of absorption is responsible for a faster rate of absorption in the plasma, but the effect of the rate constant of elimination is still far more sensible. As a result, diltiazem HCl, with the lowest rate constant of absorption and the largest rate constant of elimination, exhibits the lowest peak attained at the longest time, whatever the dosage form used for its release.
- From the considerations made previously, it appears that it is necessary to design a controlled-release dosage form with the appropriate shape and size for each drug in order to achieve the desired plasma drug level.

- Moreover, the interest in keeping the controlled-release dosage form within the GIT over a period of time much longer than the usual GIT time becomes apparent, by comparing the plasma drug levels shown in the curves noted 2' and 3' for each drug.
- Various parameters already appear to play an important role on the plasma drug level, with the nature of the drug itself, but also with the nature of the polymeric system capable of releasing the drug slowly, as well as the shape of the dosage form and its size.
- This method can calculate not only the plasma drug level but also the kinetics of release and of elimination, whatever the shape of the dosage form and the nature of the drug, provided that the pharmacokinetic parameters of the drug and the parameters of diffusion of the polymeric system are known or previously determined. Thus, this method of calculation can predict the nature of the polymeric system through which the drug should be dispersed, as well as the shape and size of the dosage form to build, in order to meet the requirements of the desired therapy. Moreover, the parameters necessary for this calculation are shown exactly, and the desirable knowledge and the methods needed for these measurements will be described at the end of this chapter.

Of course, the pharmacokinetic parameters may vary during the process. For instance, the rate constant of absorption could vary along the GIT, as suggested in an interesting paper [12]. It should be noted that the method of calculation described here is adaptable to every modification in the course of the process without losing its reliability. The main problem is to acquire the pharmacokinetic knowledge as precisely as possible.

Because the plasma drug profile stands for the most attractive information (in the first approach at least), the values of the time at which the concentration of the drug has fallen to half its maximum at the peak may be of great concern, in the same way as the half-time for a drug delivered via bolus i.v. Thus, the values of these times $T_{0.5}$ and $t_{0.5}$ are collected in Table 6.1 for these three drugs, whose plasma drug levels are drawn in Figure 6.2, Figure 6.3, and Figure 6.4 (curves 1', 2', and 3'). From these curves and the values of the half-time shown in Table 6.1, it clearly appears that, by selecting either the right polymer matrix or the dimensions, it is possible to obtain the desirable plasma drug profile for the drug, whatever its pharmacokinetic parameters, and especially its half-life time resulting from the rate of elimination.

TABLE 6.1
Half-Life Time of Various Drugs in Various Dosage Forms (h)

Drug	$t_{0.5}$	$T_{0.5}$(Immediate Release)	$T_{0.5}$	(Sustained Release)
Theophylline	5	6.5	13	20.7
Caffeine	4	5.4	11.5	17.5
Diltiazem HCl	3	6	10	15.3

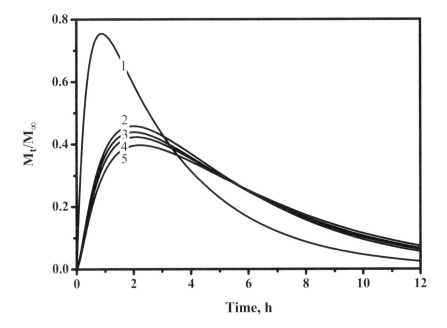

FIGURE 6.5 Plasma drug profiles with dosage forms of same volume and various shapes. 1. Immediate release. 2. Parallelepiped with $2a = 2b = c$. 3. Cube. 4. Cylinder with $R = H$. 5. Sphere with $R = 0.182$ cm. $D = 2.5 \times 10^{-7}$ cm^2/s. Aspirin with $k_a = 2.77$/h and $k_e = 0.32$/h.

6.2.2 EFFECT OF THE SHAPE OF THE DOSAGE FORM

The effect of the shape given to the dosage forms is considered in Figure 6.5, where the plasma drug level has been calculated for various dosage forms of the same volume and different shapes. The same volume specifies that there is the same amount of drug in every dosage form and the concentration of the drug is similar in each polymeric system. The following shapes are examined in succession:

1. The drug level procured by the immediate release (for purposes of comparison with the four following plasma drug levels obtained when the process of release is controlled by diffusion)
2. Parallelepiped with $2a = 2b = c$
3. Cube
4. Cylinder with $R = H$
5. Sphere of radius $R = 0.182$ cm

From these curves, some remarks of interest are made:

- The plasma drug levels are far more constant with the released dosage forms than with their immediate-release counterparts, with lower peaks and higher concentration after 12 h of release, which is a trough when these dosage forms are taken twice a day.

TABLE 6.2
Half-Life of a Drug in Dosage Forms
of Various Shapes and the Same Volume

Drug Delivery (h)	$t_{0.5}$	$T_{0.5}$
Bolus i.v.	2.1	
Immediate release	(1)	3.3
Long parallelepiped	(2)	6.2
Cube	(3)	6.7
Cylinder with $R = H$	(4)	7
Sphere	(5)	7.3

- The shapes given to the dosage forms of the same volume play a minor role. Nevertheless, it can be said that the more constant drug level is acquired with the sphere (curve 5), with a lower peak and a larger trough. A similar result has already been obtained in calculating the kinetics of drug release with these same dosage forms in Chapter 4 (Section 4.1.3, shown in Figure 4.4).
- With this variety of Aspegic having a rather high half-life (as compared with acetyl salicylic acid), the values collected in Table 6.2 provide a more precise definition of all the curves drawn in Figure 6.5, which look similar. Nevertheless, it clearly appears that the values of the half-life of these sustained-release dosage forms are within a narrow range, not differing very much from one another.

6.2.3 EFFECT OF THE NATURE OF THE DRUG

Even in the dosage form with controlled release, the drug exhibits a behavior of its own, playing a typical role. As already shown in Chapter 3 (Section 3.2.1 and Section 3.2.2), the effect of the rate constant of absorption on the plasma level is not significant, but this is not the case in the rate constant of elimination. As depicted in Figure 6.6, which shows the plasma drug levels obtained with the drugs theophylline (1), caffeine (2), and diltiazem (3), the following curves are built, with the y-axis being the amount of drug at time t in the blood as a fraction of the amount of drug located in the dosage form and the x-axis being the time (in hours), for dosage forms spherical in shape of the same radius made of the same polymer matrix and exhibiting the same diffusivity. The following conclusions can be traced:

- The drug concentration in the plasma is not shown, because it depends on the nature of the drug, with its typical metabolism and volume of distribution. Nevertheless, these curves are similar to the blood concentration profiles, because the plasma concentration is proportional to the amount of drug in the blood.
- The same polymer matrix and dimensions of the dosage form have been selected for calculation, making comparison of the three drugs easy.

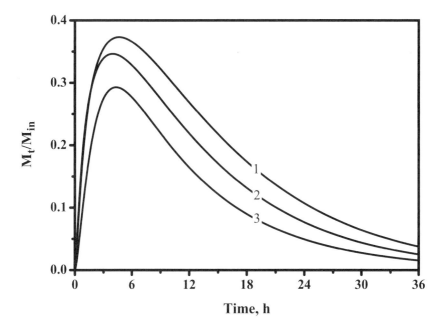

FIGURE 6.6 Plasma drug profiles with three drugs in similar sustained-release dosage forms, spherical in shape, made of the same polymer with the same diffusivity. $R = 174\,\mu m$ $D = 8.2 \times 10^{-10}\,cm^2/s$. 1. Theophylline with $k_a = 2.5/h$ and $k_e = 0.14/h$. 2. Caffeine with $k_a = 3.03/h$ and $k_e = 0.17/h$. 3. Diltiazem HCl with $k_a = 0.96/h$ and $k_e = 0.23/h$.

- The rate at which the drug emerges in the blood compartment is very high for the three drugs, and much higher for drug 1 (theophylline) and drug 2 (caffeine), which have a higher rate constant of absorption than the third (diltiazem).
- The amount of drug at the peak (or the concentration) is larger for drug 1, which has the lowest rate constant of elimination. The well-known statement holds: the higher the rate constant of elimination, the lower the amount of drug at the peak.
- The half-life time obtained with this same polymer matrix varies with the nature of the drug and the following statement holds true: for the same polymer matrix that imposes the same diffusivity, the higher the half-life time of the drug in bolus i.v., the higher the half-life time with the sustained-release dosage form.

6.2.4 EFFECT OF THE POLYMER ON THE SUSTAINED RELEASE

The polymer plays a major role in the sustained-release process through diffusivity, which measures the rate at which the polymer allows the drug to diffuse out of the dosage form. In fact, the dimension also is a parameter able to be considered, because the time of release is proportional to the square of the smaller dimension (e.g., the radius of a sphere). Nevertheless, some limits exist:

- The radius of a sphere cannot reasonably exceed 0.5 cm.
- The diffusivity as a fraction of the square of the radius (or the main dimension) of the dosage form cannot be lower than 10^{-6}/s, because such a low diffusivity is responsible for too low a rate of diffusion, leading to the drawbacks that the whole drug initially in the dosage form cannot be released and the plasma drug concentration cannot attain an efficient level over a reasonable diffusion time of one day.
- The GIT time largely varies, depending on the food taken by the patient, but cannot be generally larger than one day—or more often half a day.

Various techniques have been developed to prolong the release time of the drug from the dosage form (and thus the time over which the dosage form stays in the GI). These methods are described in various reviews [13–16]. Essentially, they consist of increasing the residence time in the GI, and more specifically, using dosage forms that float in the stomach [17]. Various studies have been carried out for the purpose of evaluating the efficacy of the techniques by comparing the kinetics of drug release either through in vitro or through in vivo experiments [18–22].

The effects of the nature of the polymer and the main dimension of the dosage form are studied by determining the plasma drug profiles obtained with various values of the ratio $\frac{D}{R^2}$ expressed in second^{-1}. Various drugs with quite different values of the pharmacokinetic parameters are considered in succession. Thus, the plasma drug profiles are drawn by plotting the amount of drug in the plasma as a fraction of the amount of drug initially in the dosage form as a function of time, for various values of this ratio $\frac{D}{R^2}$: in Figure 6.7 for ciprofloxacin, in Figure 6.8 for acetyl salicylic acid, and in Figure 6.9 for cimetidine (curves 1, 2, 3). In the same way, the kinetics of drug release are shown for the three values of the ratio (curves 1', 2', 3'). The curves in these three figures lead to the following conclusions:

- As shown in the previous section (Section 6.2.2), the effect of the drug on its release remains of great importance, whatever the polymer matrix and the main dimension of the dosage form (a sphere in the present case), and then, the ratio $\frac{D}{R^2}$, which combines them. Thus, quite different values of this ratio must be determined to prolong the plasma drug profile over a period of 24 h for each drug.
- In the special case of ciprofloxacin, with a low rate constant of elimination, the values of $\frac{D}{R^2}$ found of interest are between 10^{-5} and 5×10^{-6}/s, as shown in Figure 6.7. It should be noted that the lower value of 10^{-6}/s used in this figure leads to a kinetics (curve 1') so slow that only 75% of the drug initially in the dosage form is released after one day.
- For acetyl salicylic acid, whose rate constant of elimination is as high as 2.1/h, much lower values of this ratio would be necessary to extend the plasma drug profile over 24 h. But as already said, it is not possible to use a polymer matrix with ratio $\frac{D}{R^2}$ lower than 5×10^{-6}/s, for the same

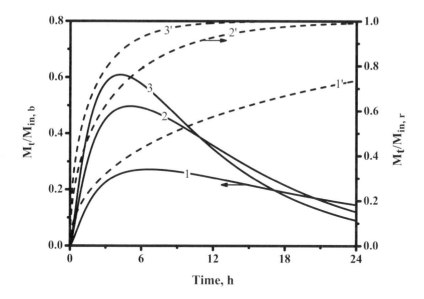

FIGURE 6.7 Plasma drug profiles (1, 2, 3) and kinetics of drug release (1', 2', 3') obtained with ciprofloxacin dispersed into three various sustained release dosage forms, spherical in shape, characterized by the ratio of the diffusivity and the square of the radius: $\frac{D}{R^2} = 10^{-5}$/s (3); $= 0.5 \times 10^{-5}$/s (2); $= 10^{-6}$/s (1). $k_a = 1.3$/h and $k_e = 0.12$/h.

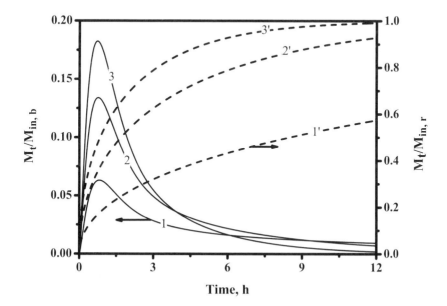

FIGURE 6.8 Plasma drug profiles (1, 2, 3) and kinetics of drug release (1', 2', 3') obtained with acetyl salicylic acid released from three various sustained-release dosage forms, spherical in shape, characterized by the ratio of the diffusivity and the square of the radius $\frac{D}{R^2} = 10^{-5}$/s (3); $= 0.5 \times 10^{-5}$/s (2); $= 10^{-6}$/s (1). $k_a = 3.5$/h and $k_e = 2.1$/h.

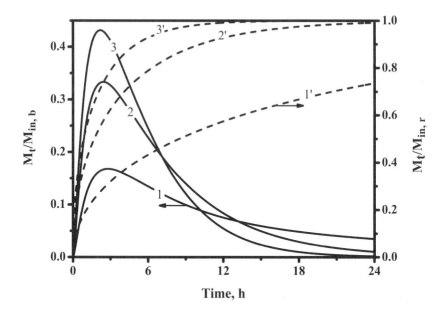

FIGURE 6.9 Plasma drug profiles (1, 2, 3) and kinetics of drug release (1', 2', 3') obtained with cimetidine released from three various sustained-release dosage forms, spherical in shape, characterized by the ratio of the diffusivity and the square of the radius $\frac{D}{R^2} = 10^{-5}$/s (3); = 0.5×10^{-5}/s (2); = 10^{-6}/s (1). $k_a = 2.2$/h and $k_e = 0.34$/h.

reasons as for ciprofloxacin. With cimetidine having a rate constant of elimination of 0.34/h, the value of the ratio $\frac{D}{R^2}$ equal to 5×10^{-6}/s can be used to satisfy the same requirement.

- A slight change in the values of $\frac{D}{R^2}$ is highly sensible for the plasma drug profiles. So, for ciprofloxacin, using the values of either 10^{-5} or 5×10^{-6}/s gives rise to a change by 25% in the maximum value at the peak, whereas the corresponding values of the half-life time $T_{0.5}$ vary from 13.2 to 16.4 h, respectively, as shown in Figure 6.7.

- It should be noted that for a drug like acetyl salicylic acid, having the large rate of elimination of 2.1/h, it is not possible to define the characteristics of the dosage form able to extend the time of release over one day for three reasons. As proved in Figure 6.8, a special floating system would be necessary to maintain the dosage form in the stomach over a period of time exceeding one day. Moreover, because of the very low value of the ratio $\frac{D}{R^2}$ and because the radius of the dosage form is limited by the value of 0.5 cm (which is the upper limit for an oral dosage form), a diffusivity as low as 10^{-6} cm^2/s is responsible for such a slow release rate that the plasma drug level is very low (curve 1).

- A relationship between the half-life time of the drug obtained either with bolus i.v. or with a sustained-release dosage form may be of help, by using various appropriate values of the ratio $\frac{D}{R^2}$ expressed in second^{-1}, as shown later in Figure 6.26 and Figure 6.27.

TABLE 6.3
Pharmacokinetic Parameters of the Drugs

Drug	Ciprofloxacin [23]			Acetyl Salicylic Acid [24]			Cimetidine [25, 26]		
k_a (/h)	1.3			3.5			2.2		
k_e (/h)	0.09	0.12	0.22	2.1	2.5	2.77	0.27	0.34	0.57
$t_{0.5}$ (h)	7.7	5.7	3.1	0.33	0.28	0.25	2.6	2	1.2
$T'_{0.5}$ (h)	10.5	8.4	5.7						
$T_{0.5}$ (h)	19.8	16.51	11.1	2.28	2.08	1.97	9.2	7.5	5.7
$\frac{M_{max}}{M_{in}}$	0.53	0.49	0.39	0.13	0.12	0.11	0.37	0.33	0.26

6.2.5 EFFECT OF PATIENTS' INTERVARIABILITY

The effect of the intervariability of the patients is evaluated by considering the mean and the extreme values of the pharmacokinetic parameters found in the literature for the drugs—among them, the rate constant of elimination is the best-known and the most sensible. Three drugs—ciprofloxacin, acetyl salicylic acid and cimetidine—were selected and the values of their rate constant of elimination are given in Table 6.3, as well as their extreme values, taking patients' intervariability into account. It is worth noting that the values of the rate constant of elimination for ciprofloxacin are selected from the highest found in the literature, as shown in Table 2.1 (Chapter 2).

The plasma drug profiles are calculated with sustained-release dosage forms characterized by the value of the ratio $\frac{D}{R^2}$ equal to 5×10^{-6}/s and drawn in Figure 6.10 for ciprofloxacin, Figure 6.11 for acetyl salicylic acid, and Figure 6.12 for cimetidine. For each drug, the same value of this ratio $\frac{D}{R^2}$ is used, as well as the three values of the rate constant of elimination shown in Table 6.3.

From the observation of these curves and of the values collected in Table 6.3, the following comments are worth making:

- As shown previously (Section 6.2.4), each drug should be dispersed through a dosage form with a radius (or the mean dimension for a nonspherical dosage form) and made of a polymer matrix exhibiting a typical diffusivity, so that the ratio $\frac{D}{R^2}$ is slightly larger than the minimum value of 10^{-6}/s.
- The polymer matrix of the dosage form can prolong the time over which the drug is delivered, but the increase in the half-life time $T_{0.5}$ obtained with the sustained-release dosage form also depends on the half-life time $t_{0.5}$ of the drug delivered through bolus i.v. Thus, in the case of ciprofloxacin, for the lower value of $t_{0.5}$, the corresponding value of $T_{0.5}$ is multiplied by nearly 4, whereas it is multiplied by only 2.5 for the larger value of 7.7 h. The half-life times allowed by using this sustained-release system are about twice those obtained with the immediate-release dosage forms (noted $T'_{0.5}$ in Table 6.3) for the same drug.
- Intervariability exists not only when the drug is delivered through bolus i.v. but also in the case of oral sustained-release dosage forms. Nevertheless, if we consider the ratio of the larger value for $T_{0.5}$ to the lower one with the

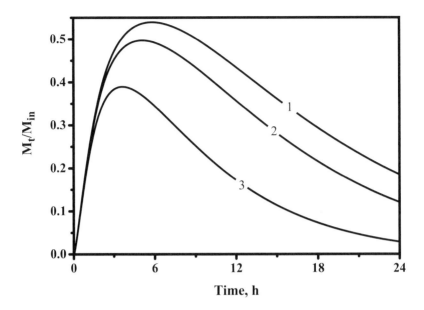

FIGURE 6.10 Effect of patients' intervariability on the plasma drug profiles obtained for ciprofloxacin in the sustained release dosage forms with $\frac{D}{R^2} = 5 \times 10^{-6}/s$. $k_a = 1.3/h$ and, respectively, 1. $k_e = 0.09/h$, 2. $k_e = 0.12/h$, 3. $k_e = 0.22/h$.

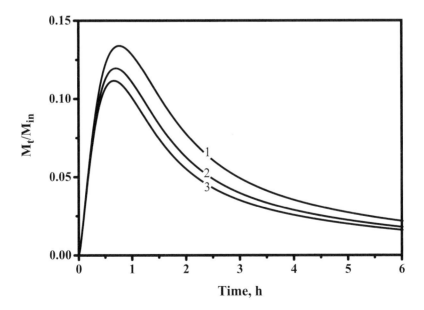

FIGURE 6.11 Effect of patients' intervariability on the plasma drug profiles obtained for acetyl salicylic acid in the sustained-release dosage forms with $\frac{D}{R^2} = 5 \times 10^{-6}/s$. $k_a = 3.5/h$ and, respectively, 1. $k_e = 2.1/h$, 2. $k_e = 2.5/h$, 3. $k_e = 2.77/h$.

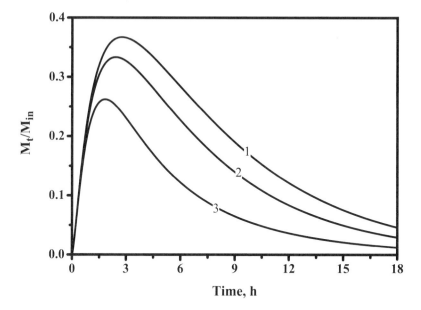

FIGURE 6.12 Effect of patients' intervariability on the plasma drug profiles obtained for cimetidine in the sustained-release dosage forms with $= 5 \times 10^{-6} \, cm^2/s \, k_a = 2.5/h$ and, respectively, 1. $k_e = 0.27/h$, 2. $k_e = 0.34/h$, 3. $k_e = 0.57/h$.

sustained-release dosage form, it is found to be 1.8 for ciprofloxacin, whereas the same ratio is 2.5 for the half-life times obtained through bolus i.v. The same fact can be observed for acetyl salicylic acid (the ratios of the half-life times are, respectively, 1.16 and 1.32) and for cimetidine (for which the ratios are 1.6 and 2.2). Thus, we can expect a far more constant plasma drug profile over a long period of time with repeated doses, when they are delivered with the appropriate sustained-release dosage forms, than with either bolus i.v. or oral immediate-release dosage forms.

• The variation in the concentration at the peaks (measured at the maximum value of the ratio $\frac{M_{max}}{M_{in}}$) could also be of interest. These values, collected in Table 6.3, show that the effect of patients' intervariability is of great concern, whatever the system of drug delivery, because the ratios of these maximum values measured in the extreme cases of the rate constant of elimination are 1.36, 1.18, and 1.42, respectively, for these three drugs when they are dispersed through a polymer matrix characterized by the ratio of 5×10^{-6}/s.

6.3 SUSTAINED RELEASE WITH REPEATED DOSES

6.3.1 EXAMPLES WITH VARIOUS DRUG–POLYMER COUPLES

The three dosage forms whose release has been previously widely studied when delivered with one dose (Section 6.2.1) are now considered in multidosing taken once a day. Thus, the plasma drug profile is drawn for theophylline (Figure 6.13),

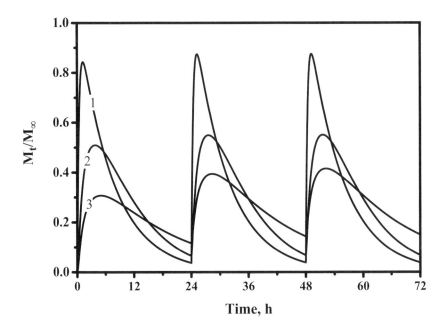

FIGURE 6.13 Plasma drug concentration with theophylline dispersed through three dosage forms, taken once a day. 1. Immediate release. Sustained release from spherical dosage forms with 2. $R = 87$ μm and 3. with $R = 174$ μm. $D = 5 \times 10^{-6}$ cm²/s; $k_a = 2.5$/h; $k_e = 0.14$/h.

caffeine (Figure 6.14), and diltiazem HCl (Figure 6.15) by plotting the amount of drug in the blood as a fraction of the amount of drug initially in the dosage form, as a function of time.

In all three figures, the notation is as follows:

- **Curve 1:** Drug profiles obtained with immediate release
- **Curve 2:** Controlled release, with 85% of the drug released within 6 h
- **Curve 3:** Controlled release, with 85% of the drug released within 24 h

The following conclusions can be drawn:

- The advantage of controlled-release systems over their immediate-release counterparts appears clearly for every drug: The dosage forms with controlled release are responsible for more constant plasma drug levels, with lower peaks and higher troughs.
- The values of the ratio $\frac{D}{R^2}$ in all cases is between 5×10^{-6} and 2×10^{-7}/s, whatever the shape of the dosage form. Of course, the dosage forms with larger dimensions and lower value of this ratio (or its equivalent), drawn in curve 3, exhibit the most constant plasma drug level but also the lowest one.
- The effect of the nature of the drug is also clearly apparent when the plasma drug levels in Figure 6.13, Figure 6.14, and Figure 6.15 are compared. Briefly, diltiazem HCl (Figure 6.15), with the lowest rate constant of absorption and

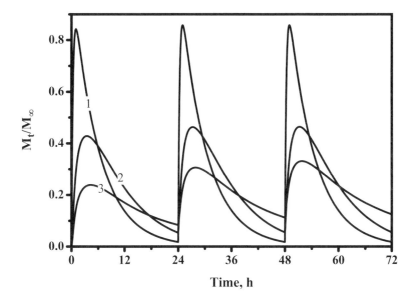

FIGURE 6.14 Plasma drug concentration with caffeine dispersed through three dosage forms, taken once a day. 1. Immediate release. Sustained release from cylindrical dosage forms with 2. $R = 112$ μm and $H = 185$ μm and with 3. $R = 250$ μm and $H = 320$ μm. $D = 8.2 \times 10^{-10}$ cm^2/s; $k_a = 3.03$/h; $k_e = 0.17$/h.

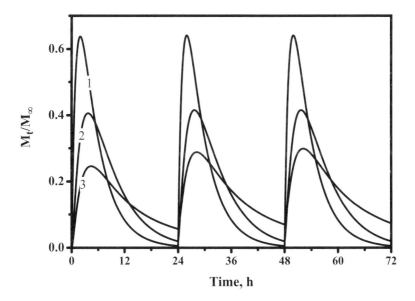

FIGURE 6.15 Plasma drug concentration with diltiazem HCl dispersed through three dosage forms, taken once a day. 1. Immediate release. Sustained release from parallelepiped dosage forms with 2. $96 \times 96 \times 192$ μm and with 3. $192 \times 192 \times 384$ μm. $D = 2.5 \times 10^{-10}$ cm^2/s; $k_a = 0.96$/h; $k_e = 0.23$/h.

the largest rate constant of elimination, exhibits the lowest plasma drug level with lowest peaks and lowest troughs, whereas the profiles of the other two drugs (drawn in Figure 6.13 and Figure 6.14) are nearly identical.

- Of course, the so-called steady state (meaning that the profiles are similar for each dose taken in succession) is only attained when the trough associated with the previous dose is very low. Thus for curve 3, obtained with the largest dimensions in these three figures, the steady state is attained only after the third dose, whereas it is attained after the second dose in profile 2 and at the first dose with immediate-release dosage form, for which the trough is very low after 24 h.

6.3.2 EFFECT OF THE NATURE OF THE DRUG ON THE PROFILES

The plasma drug profiles are drawn in Figure 6.16 for the three drugs, dispersed through similar sustained-release dosage forms, spherical in shape, made of the same polymer with the same diffusivity, and taken once a day:

1. Theophylline
2. Caffeine
3. Diltiazem HCl

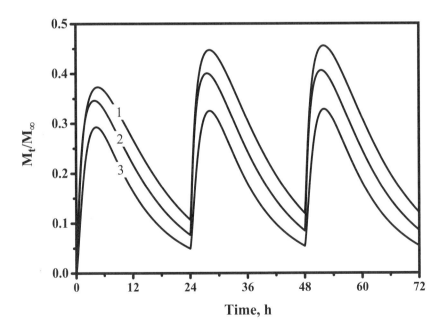

FIGURE 6.16 Plasma drug profiles with three drugs in similar sustained-release dosage forms, spherical in shape, made of the same polymer with the same diffusivity, taken once a day. $R = 174$ μm $D = 8.2 \times 10^{-10}$ cm^2/s $\frac{D}{R^2} = 2.7 \times 10^{-6}$ s. 1. Theophylline with $k_a = 2.5$/h and $k_e = 0.14$/h. 2. Caffeine with $k_a = 3.03$/h and $k_e = 0.17$/h 3. Diltiazem HCl with $k_a = 0.96$/h and $k_e = 0.23$/h.

Because these dosage forms have the same characteristics, comparing them is easy, as previously shown for a single dose in Figure 6.6 (Section 6.2.3). The curves with repeated doses in Figure 6.16 lead to the following conclusions:

- The plasma drug level in Figure 6.16 is not as low as in the previous figures (6.13, 6.14, and 6.15), because the new value selected for the ratio $\frac{D}{R^2} = 2.7 \times 10^{-6}$/s is larger for the three drugs.
- The results are somewhat similar to those found for the single dose shown in Figure 6.6, at least for the values of the peaks and of the half-life times $T_{0.5}$. Thus, the highest peak and the longest half-life time are attained with theophylline (curve 1), and the lowest ones are obtained with diltiazem HCl (curve 3).
- The lowest trough is attained for diltiazem HCl, which has the largest rate constant of elimination, and the highest trough is obtained for theophylline, which has the lowest rate constant of elimination. Following this result, the so-called steady state can be observed after the second dose for theophylline and already after the second dose for the other two drugs.

6.3.3 EFFECT OF THE POLYMER MATRIX ON THE PROFILES

As already noted for a single dose (Section 6.2.4), the polymer matrix through which the drug is dispersed plays an important role on the sustained release of the drug, when the dosage forms are taken repeatedly. The two main characteristics of the dosage form (e.g., the diffusivity of the polymer and the main dimension, which is the radius in the case of a sphere) are then expressed by the ratio $\frac{D}{R^2}$. Thus, the following values of this ratio are used in calculating the plasma drug level for these three drugs: ciprofloxacin (Figure 6.17), acetyl salicylic acid (Figure 6.18), and cimetidine (Figure 6.19).

From these curves, the following observations are worth noting:

- For each drug, the effect of the value given to the ratio $\frac{D}{R^2}$ is of great concern, because it acts upon the plasma drug level. The highest peaks associated with the lowest troughs are obtained with the larger value of this ratio (curve 3); thus, generally, decreasing the value of this ratio also provokes a significant decrease in the concentration at the peak and a major relative increase at the trough.
- The effect of the nature of the drug, through the value of its rate constant of elimination, remains important, as already shown for a single dose (Section 6.2.3). For repeated doses, it is possible to prepare convenient sustained-release dosage forms able to be taken once a day for ciprofloxacin, whose rate constant of elimination is very low. However, for acetyl salicylic acid, the same dosage forms can be taken only four times a day (every 6 h) because of its large value of the rate constant of elimination. For cimetidine, with an intermediate value of 0.34/h of the rate constant of elimination, these dosage forms can be taken twice a day.

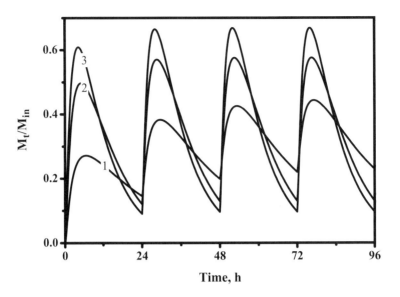

FIGURE 6.17 Plasma drug profiles (1, 2, 3) with repeated doses taken once a day, obtained with ciprofloxacin dispersed into three various sustained-release dosage forms, spherical in shape, characterized by the ratio of the diffusivity and the square of the radius: $\frac{D}{R^2} = 10^{-5}$/s (3); $= 0.5 \times 10^{-5}$/s (2); $= 10^{-6}$/s (1). $k_a = 1.3$/h and $k_e = 0.12$/h.

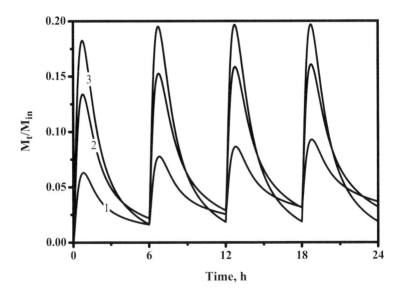

FIGURE 6.18 Plasma drug profiles (1, 2, 3) with repeated doses taken four times a day, obtained with acetyl salicylic acid released from three various sustained-release dosage forms, spherical in shape, characterized by the ratio of the diffusivity and the square of the radius $\frac{D}{R^2} = 10^{-5}$/s (3); $= 0.5 \times 10^{-5}$/s (2); $= 10^{-6}$/s (1). $k_a = 3.5$/h and $k_e = 2.1$/h.

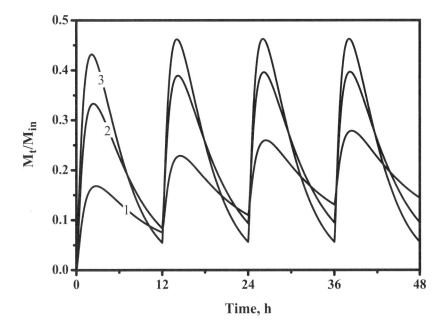

FIGURE 6.19 Plasma drug profiles (1, 2, 3) with repeated doses taken twice a day, obtained with cimetidine released from three various sustained-release dosage forms, spherical in shape, characterized by the ratio of the diffusivity and the square of the radius $\frac{D}{R^2} = 10^{-5}$/s (3); $= 0.5 \times 10^{-5}$/s (2); $= 10^{-6}$/s (1). $k_a = 2.2$/h and $k_e = 0.34$/h.

- The effect of the nature of the drug also acts upon the value at the peak, a lower value of the rate of elimination being responsible for a higher peak. Thus, the values at the peak decrease in this order: ciprofloxacin > cimetidine > acetyl salicylic acid.
- The so-called steady state is attained rapidly when the concentration at the trough of the previous dose is very low. Thus, it is attained after the second dose for the two largest values of the ratio $\frac{D}{R^2}$, whereas it is obtained only at the fourth dose for the lower value of this ratio, under these dosage conditions:
  - Once a day for ciprofloxacin
  - Four times a day for acetyl salicylic acid
  - Twice a day for cimetidine

6.3.4 Effect of the Patients' Intervariability

The effect of the intervariability of the patients on plasma drug profiles is determined by considering the following three drugs, when they are dispersed through a spherical matrix and taken in repeated doses:

- Ciprofloxacin (Figure 6.20)
- Acetyl salicylic acid (Figure 6.21)
- Cimetidine (Figure 6.22)

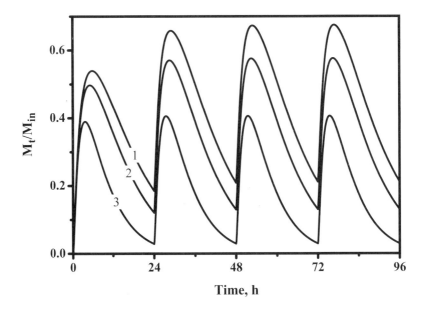

FIGURE 6.20 Effect of patients' intervariability on the plasma drug profiles obtained for ciprofloxacin in the sustained-release dosage forms taken once a day with $\frac{D}{R^2} = 5 \times 10^{-6}$/s $k_a = 1.3$/h and, respectively, 1. $k_e = 0.09$/h, 2. $k_e = 0.12$/h, 3. $k_e = 0.22$/h.

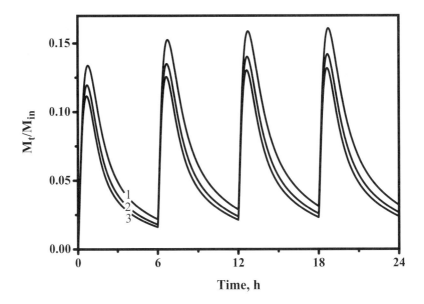

FIGURE 6.21 Effect of patients' intervariability on the plasma drug profiles obtained for acetyl salicylic acid in the sustained-release dosage forms taken four times a day, with $\frac{D}{R^2} = 5 \times 10^{-6}$/s $k_a = 3.5$/h and, respectively, 1. $k_e = 2.1$/h, 2. $k_e = 2.5$/h, 3. $k_e = 2.77$/h.

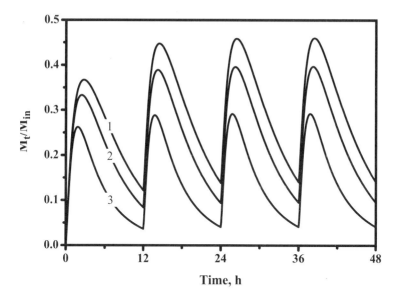

FIGURE 6.22 Effect of patients' intervariability on the plasma drug profiles obtained for cimetidine in the sustained-release dosage forms, taken twice a day with $\frac{D}{R^2} = 5 \times 10^{-6}$/s and, respectively, $k_a = 2.2$/h, 1. $k_e = 0.27$/h, 2. $k_e = 0.34$/h 0^{-6}, 3. $k_e = 0.57$/h.

The optimum characteristics of the polymer matrix selected previously (Section 6.3.3) are expressed in terms of the ratio $\frac{D}{R^2} = 5 \times 10^{-6}$/s. The various values of the rate constant of elimination are shown in Table 6.3 for these three drugs.

A few comments can be made based on observation of the curves in Figure 6.20 through Figure 6.22:

- Just as with a single dose (Section 6.2.5 and Figure 6.10 through Figure 6.12), patients' intervariability should be considered seriously. For all three drugs, the change in the value of the rate constant of elimination is the factor of concern. Of course, the following statement holds: the lower the rate constant of elimination, the larger the plasma drug level with a higher peak attained at a longer time.
- One factor that intervenes in repeated doses is the drug concentration at the trough. When the drug level at the trough is high, the steady state requires a larger number of doses to be reached. Thus, the steady state is attained in all cases at the third dose, except for ciprofloxacin and cimetidine, because of their larger rate constant of elimination, as shown in Figure 6.20 (curve 3) and in Figure 6.22 (curve 3), respectively, for which it has already been attained at the second dose.

6.3.5 Effect of the Dose Intervals (Frequency)

With repeated doses, the factor of concern is the intervals between doses, or dose frequency, for the purpose of coming to a more constant plasma drug level. In order

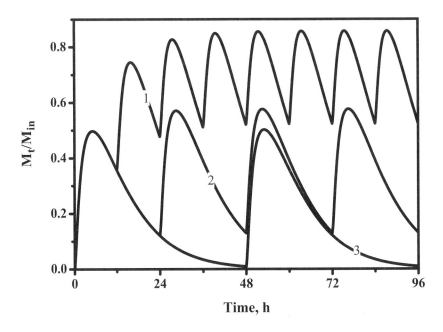

FIGURE 6.23 Plasma drug profiles obtained with ciprofloxacin in the sustained-release dosage form with $\frac{D}{R^2} = 5 \times 10^{-6}$/s, when taken 1. twice a day 2. once a day, and 3. every other day. $k_a = 1.3$/h and $k_e = 0.12$/h.

to study this parameter precisely, the plasma drug level is determined in considering the three following drugs:

- Ciprofloxacin (Figure 6.23)
- Acetyl salicylic acid (Figure 6.24)
- Cimetidine (Figure 6.25)

The main parameters characterizing sustained release are the same for these three drugs, with the ratio $\frac{D}{R^2} = 5 \times 10^{-6}$/s. Some conclusions can be drawn from these curves:

- The effect of the nature of the drug on the plasma drug level clearly appears with these three drug profiles, and the rate constant of elimination of the drug remains the main parameter. Thus, in order to obtain a constant plasma drug level, with ciprofloxacin (Figure 6.23) whose constant rate constant of elimination is no more than 0.12/h, it is possible to prepare some dosage forms that would be taken twice a day or even once a day. In contrast, for acetyl salicylic acid (Figure 6.24), which has a high rate constant of elimination (2.1/h), the dosage forms would be taken four times a day, and for cimetidine (Figure 6.25), whose rate constant rate of elimination is low (0.34/h) but not as low as that of ciprofloxacin, the dosage forms would be taken twice a day.
- Another way to express the same result is to consider the effect of the dose interval (frequency) on the plasma profile for these three drugs.

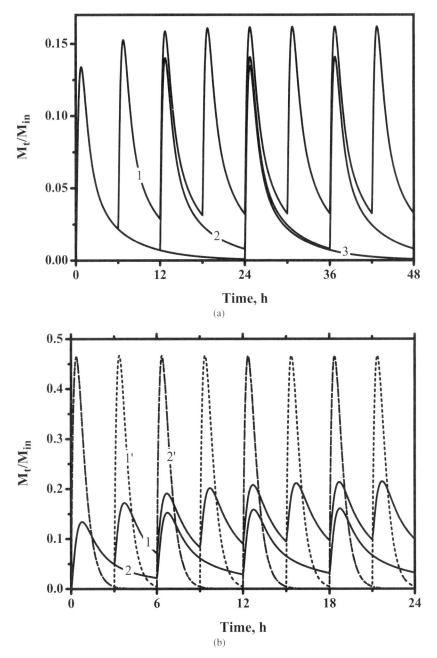

FIGURE 6.24 (a) Plasma drug profiles obtained with acetyl salicylic acid in the sustained-release dosage form with $\frac{D}{R^2} = 5 \times 10^{-6}$/s, when taken 1. four times a day, 2. twice a day, and 3. once a day. $k_a = 3.5$/h and $k_e = 2.1$/h; (b) Plasma drug profiles obtained with acetyl salicylic acid in the sustained-release dosage form with $\frac{D}{R^2} = 5 \times 10^{-6}$/s, when taken every 3 h (solid line 1) or every 6 h (solid line 2), and obtained with immediate release taken every 3 h (dotted line 1') or every 6 h (dotted line 2'). $k_a = 3.5$/h and $k_e = 2.1$/h.

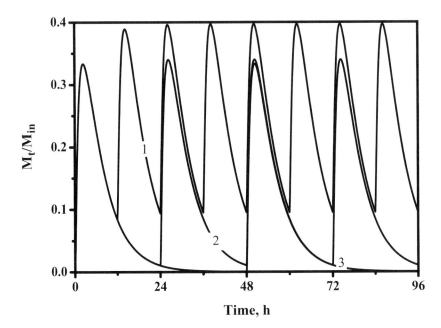

FIGURE 6.25 Plasma drug profiles obtained with cimetidine in the sustained-release dosage form with $\frac{D}{R^2} = 5 \times 10^{-6}$/s, when taken 1. twice a day, 2. once a day, and 3. every other day. $k_a = 3.5$/h and $k_e = 2.$l/h.

Ciprofloxacin, when taken once a day in this sustained-release dosage form, exhibits a rather flat plasma drug level with high peaks and low troughs, the steady state being attained from the second dose. The same drug in the same dosage form, when taken twice a day, leads to a quite different plasma drug profile, because there is an undulating concentration pattern of the drug in the blood in which high peaks alternate with high troughs, making it possible to maintain the concentration at the optimal therapeutic level when the steady state is acquired after the fourth dose.

In sharp contrast with ciprofloxacin, acetyl salicylic acid (Figure 6.24) exhibits a typical pattern for its plasma drug profile—very acute peaks alternating with very low troughs—so that the optimal therapeutic level is only briefly present when the dosage form is taken twice a day or even four times a day. In fact, a drug with such a high rate of elimination necessitates the use of gastroretention drug delivery systems based on either floating or sticking extended-release delivery processes [13–16]. Comparison can be made in Figure 6.24b, which shows the plasma drug profiles calculated with the drug delivered with immediate release (dotted line) and with sustained-release dosage forms with the ratio $\frac{D}{R^2} = 5 \times 10^{-6}$/s.

Although the half-life time is only 1 h for the immediate-release dosage form, it attains 2.33 h with the sustained-release dosage form. Thus with such a sustained-release dosage form, it is possible to obtain a plasma drug profile of interest when the dose interval is between 3 and 6 h, the steady state being attained after the fourth dose and the third dose in each case, respectively. It should be noted that the amount of drug in the blood is very low because of the high rate of elimination. The plasma drug profiles calculated with the immediate-release dosage form are similar whether the dosage interval is 3 or 6 h (dotted lines 1' and 2').

With cimetidine, whose rate constant of elimination is 0.34/h, it is possible to obtain a plasma drug profile of interest when the dose frequency is twice a day (Figure 6.25), the steady state being reached after the third dose.

6.4 PREDICTION OF THE CHARACTERISTICS OF THE DOSAGE FORM

As an extension of the results shown in this chapter, it is of interest to know exactly the main characteristics of sustained release that are necessary before building the dosage form that can obtain the desired plasma drug level.

6.4.1 RELATIONSHIPS BETWEEN HALF-LIFE TIMES

Relationships between the half-life time obtained when the drug is delivered through bolus i.v. and the half-life times obtained with various sustained-release dosage forms would be useful in predicting the right sustained-release dosage form able to comply with the desired requirement of therapy. These relationships are expressed by means of two diagrams:

- When the half-life time $t_{0.5}$ is between 1 and 10 h (Figure 6.26)
- When the half-life time $t_{0.5}$ is much lower, between 0 and 1 h (Figure 6.27)

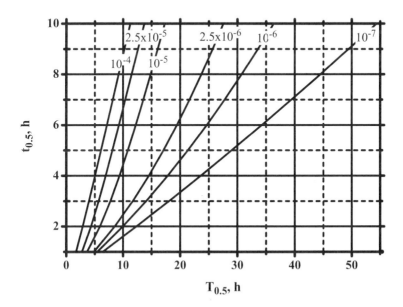

FIGURE 6.26 Diagram expressing the relation between the half-life time $t_{0.5}$ (h), obtained in bolus i.v., and the half-life time $T_{0.5}$ (h), obtained with sustained-release dosage forms, with various values of the ratio $\frac{D}{R^2}$/s, with $1 < t_{0.5} < 10$.

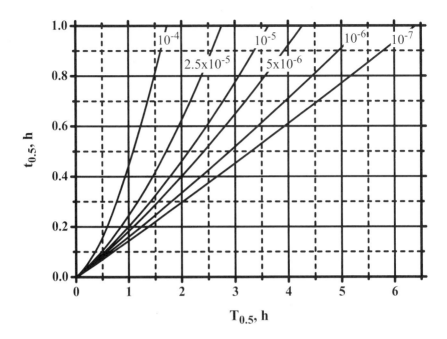

FIGURE 6.27 Diagram expressing the relation between the half-life time $t_{0.5}$ (h), obtained in bolus i.v., and the half-life time $T_{0.5}$ (h) obtained with sustained-release dosage forms, with various values of the ratio $\frac{D}{R^2}$/s, with $0 < t_{0.5} < 1$.

From the observation of these two figures, the following conclusions can be drawn:

- These diagrams are built by using various values of the ratio $\frac{D}{R^2}$ expressed in terms of s^{-1}, which combines the characteristics of the sustained-release system (e.g., the diffusivity of the polymer matrix and the main dimension of the dosage form, which is either a sphere's radius or the smaller size taken in any other shape of the dosage forms).

- The values of the ratio $\frac{D}{R^2}$/s should be larger than 10^{-7}/s, because lower values of this ratio give rise to a very low plasma drug level. The radius for the spherical dosage form cannot be larger than 0.5 cm., and the corresponding value of the diffusivity would be responsible for a very low rate of drug release from the dosage form.

- For better accuracy, especially for very low values of the half-life time obtained when the drug is delivered through bolus i.v., two diagrams are necessary to express the results: one with $1 < t_{0.5} < 10$ (Figure 6.26) and the other with $0 < t_{0.5} < 1$ (Figure 6.27).

- It clearly appears that the obvious but nevertheless fundamental statement holds: the higher the value of $t_{0.5}$, the higher the value of $T_{0.5}$ obtained with the sustained-release system. Thus, for a drug such as acetyl salicylic acid, whose half-life time through bolus i.v. is as low as 0.33 h, the larger value of the half-life time $T_{0.5}$ obtained with a sustained-release dosage form cannot be expected to be much more than 2 to 2.5 h.

6.4.2 PLASMA DRUG PROFILE IN TERMS OF CONCENTRATION

The various plasma drug profiles have been expressed in terms of time for the abscissa and by considering the amount of drug in the plasma as a fraction of the amount initially in the dosage forms for the ordinate. The interest of this presentation is that a dimensionless number appears in the ordinate, which can be used whatever the amount of drug in the dosage form. Because physicians are more sensitive to the concentration, and in addition the analysis leading directly to the drug concentration, it is necessary to be able to convert the ratio shown in preceding ordinate into the plasma drug concentration. In fact, the conversion is easy by using the apparent volume of distribution V_p (which can be very different from the plasma volume) and the amount initially dispersed into the dosage form, because there is a proportionality between the amount of drug in the plasma and the resulting concentration.

Besides the apparent volume of distribution, another pharmacokinetic parameter appears of interest: the rate constant of elimination or the half-life time, which is determined with bolus i.v. When these two parameters are known, the convertibility of the amount of drug into the drug concentration is easy to obtain, whatever the values of these two pharmacokinetic parameters (and the relation that exists between them) may be. Thus, an example is given in the case of patients' intervariability: The plasma drug profiles shown in Figure 6.10 for ciprofloxacin (Section 6.2.5),

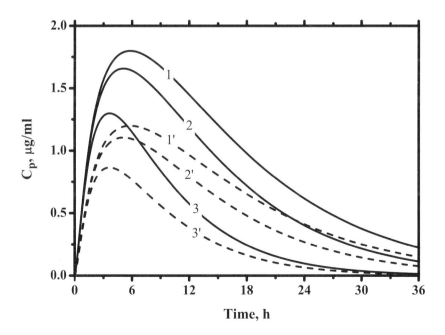

FIGURE 6.28 Plasma drug concentration of ciprofloxacin as a function of time with one dose. Patients' intervariability with the simultaneous effect of the apparent plasmatic volume and of the rate constant of elimination. $k_a = 1.3/h$ and $k_e = 0.09/h$ (1 and 1'), $k_e = 0.12/h$ (2 and 2'), and $k_e = 0.22/h$ (3 and 3') with $V_p = 150 l$ (1, 2, 3) and $V_p = 225 l$ (1', 2', 3'); $M_{in} = 500$ mg.

expressed in terms of the ratio of mass of drug $\frac{M}{M_{in}}$, are converted into plasma drug concentrations in Figure 6.28. The six curves represent the patients' intervariability by considering not only the change in the rate constant of elimination but also two extreme values of the apparent plasmatic volume (150 and 225 l).

REFERENCES

1. Nia, B., Ouriemchi, E.M., and Vergnaud, J.M., Calculation of the blood level of a drug taken orally with a diffusion controlled dosage form, *Int. J. Pharm.*, 119, 165, 1995.
2. Sournac, M. et al., Scintigraphic study of the gastro-intestinal transit and correlations with the drug absorption kinetics of a sustained release theophylline tablet, *J. Control. Release*, 7, 139, 1988.
3. Ouriemchi, E.M., Bouzon, J., and Vergnaud, J.M., Modelling the process of controlled release of drug in in vitro and in vivo tests, *Int. J. Pharm.*, 113, 231, 1995.
4. Ouriemchi, E.M. and Vergnaud, J.M., Prediction of in-vivo blood level with controlled-release dosage forms. Effect of the gastrointestinal tract time, *J. Pharm. Pharmacol.*, 48, 390, 1995.
5. Ouriemchi, E.M. and Vergnaud, J.M., Calculation of the plasma drug level with oral controlled release dosage forms: effect of the dose frequency, *Int. J. Pharm.*, 127, 177, 1996.
6. Aïnaoui, A. and Vergnaud, J.M., Modelling the plasma drug level with oral controlled release dosage forms with lipidic Gelucire, *Int. J. Pharm.*, 169, 155, 1998.
7. Ouriemchi, E.M. and Vergnaud, J.M., Plasma drug level assessment with controlled release dosage forms with a core and shell and lower concentration in the shell, *Int. J. Pharm.*, 176, 251, 1999.
8. Vergnaud, J.M., Use of polymers in pharmacy for oral dosage forms with controlled release, *Recent Res. Devel. Macromol. Res.*, 4, 173, 1999.
9. Aïnaoui, A. and Vergnaud, J.M., Effect of the nature of the polymer and of the process of drug release (diffusion or erosion) for oral dosage forms, *Comput. Theoret. Polymer Sci.*, 10, 383, 2000.
10. Aïnaoui, A. et al., Calculation of the dimensions of dosage forms with release controlled by diffusion for in vivo use, *Europ. J. Pharm. Biopharmacol.*, 51, 17, 2001.
11. Droin, A. et al., Model of matter transfer between sodium salicylate–Eudragit matrix and gastric liquid, *Int. J. Pharm.*, 27, 233, 1985.
12. Amidon, G.L. et al., A theoretical basis for a biopharmaceutic drug classification: the correlation of in vitro drug dissolution and in vivo bioavailability, *Pharm. Res.*, 12, 413, 1995.
13. Dubernet, C., Systèmes à libération gastrique prolongée, in *Nouvelles formes médicamenteuses*, Falson-Rieg, F., Faivre, V. and Pirot, F., Eds., Tec & Doc, Paris, 2004, Chapter 6.
14. Deshpande, A.A. et al., Controlled release drug delivery systems for prolonged gastric residence: an overview, *Drug Dev. Ind. Pharm.*, 22, 531, 1996.
15. Hwang, S.J., Park, H., and Park, K., Gastroretentive drug delivery systems, *Crit. Rev. Ther. Drug. Carrier Syst.*, 15, 243, 1998.
16. Dürig, T. and Fassihi, R., Evaluation of floating and sticking extended release delivery systems: an unconventional dissolution test, *J. Control. Release*, 67, 37, 2000.
17. Timmermans, J. and Moes, A.J., How well do floating dosage forms float? *Int. J. Pharm.*, 62, 207, 1990.

18. Baumgartner, S. et al., Optimisation of floating matrix tablets and evaluation of their gastric residence time, *Int. J. Pharm.*, 195, 125, 2000.
19. Ozdemir, N., Ordu, S., and Ozkan, Y., Studies on floating dosage forms of furosemide: in vitro and in vivo evaluations of bi-layer tablet formulations, *Drug Dev. Ind. Pharm.*, 26, 857, 2000.
20. Wei, Z., Yu, Z., and Bi, D., Design and evaluation of a two layer floating tablet for gastric retention using cisapride as a model drug, *Drug Dev. Ind. Pharm.*, 27, 469, 2001.
21. Rouge, N. et al., Comparative pharmacokinetic study of a floating multiple unit capsule, a high density multiple unit capsule and an immediate release tablet containing 25 mg atenolol, *Pharm. Acta Helv.*, 73, 81, 1998.
22. Whitehead, L. et al., Floating dosage forms: an in vivo study demonstrating prolonged gastric retention, *J. Control. Release,* 55, 3, 1998.
23. Fabre, D. et al., Steady-state pharmacokinetics of ciprofloxacin in plasma from patients with nosocomial pneumonia: penetration of the bronchial mucosa, *Antimicrob. Agents Chemother.,* Dec. 1991, 2521, 1991.
24. Rowland, M. and Riegelman, S., Pharmacokinetics of acetylsalicylic acid and salicylic acid after intravenous administration in man, *J. Pharm. Sci.,* 57, 1313, 1968.
25. Somogyi, A. and Roland, G., Clinical pharmacokinetics of cimetidine, *Clin. Pharmacokin.,* 8, 463, 1983.
26. Arancibia, A. et al., Pharmacocinétique de la cimetidine après administration d'une dose unique par voie intraveineuse rapide et d'une autre par voie orale, *Thérapie,* 40, 87, 1985.

7 Plasma Drug Level with Erosion-Controlled Dosage Forms

NOMENCLATURE

AUC Area under the curve.

C Plasma drug concentration.

GI(T) Gastrointestine or gastrointestinal (tract).

H Half the length of a cylinder.

i.v. Intravenous.

k_a Rate constant of absorption of the drug (per/h).

k_e Rate constant of elimination of the drug (per/h).

L Half the thickness of the sheet.

M_t Amount of drug released up to time t.

M_{in} Amount of drug initially in the dosage form.

$\frac{M_t}{M_{in}}$ Dimensionless number expressing the drug release from the dosage form.

$\frac{dM}{dt}$ Rate of drug release by the dosage form in the GI at time t.

R Radius of the sphere, or of the cylinder.

t, t_r Time, time of full erosion, respectively.

$t_{0.5}$ Half-life time of the drug delivered in bolus i.v.

$T_{0.5}$ Half-life time of the drug delivered with the sustained-release dosage form.

V_p Apparent plasmatic volume (l).

v Linear rate of erosion (cm/s).

W Amount of drug eliminated (Equation 7.3).

Y Amount of drug in the GI volume (Equation 7.1).

Z Amount of drug in the plasmatic volume (Equation 7.2).

Oral dosage forms with release controlled by erosion are prepared by dispersing the drug through an erodible polymer. With these erodible polymers, the problem of drug release can become rather complex, when the process is initiated by diffusion of the liquid into the polymer and followed by dissolution of the polymer on the surface, where the concentration of liquid reaches a high level. However, some erodible polymers exist in which the process seems to be strictly controlled by erosion [1–4].

Biodegradable polymers can be obtained in two ways:

- By using pure polymers
- By introducing some additives into the polymer that dissolve by provoking a progressive disintegration

125

An example of the first case is observed with Na carboxymethyl cellulose, the dissolution being related to the rheological properties of the gelled polymer [5]. Another example has been pointed out with Gelucires, in which the process of drug release is controlled either by diffusion with lipidic material or by erosion [1–4], depending on the hydrophilic–lipidic balance.

Moreover, the great interest in these erodible dosage forms results from the fact that they can be used in bioadhesive delivery systems; over a given period of time, it is certain that all the components of the dosage forms will be eliminated. Thus the problem of bioadhesion becomes of great concern.

It is not the purpose in this chapter to present an extensive bibliography on the subject of bioadhesion coupled with erosion. Various books and reviews have tempted to attain this objective [6, 7], and only a few examples are summarized in order to give a very brief overview of the importance of the problems encountered [8–11]. On the whole, it can be said that mucoadhesion has been related to the rheological properties of the polymer [12–14], as well as to its molecular weight by considering polydispersity or by insisting on the moisture absorption of hydroxypropyl cellulose [15, 16]. Various methods and techniques have been developed and used in order to study the mechanism of bioadhesion and its subsequent effect with the retention of the dosage form [17–21]. These various studies have attempted to answer the fundamental questions put forward concerning bioadhesion and its consequences for dosage forms with erosion-controlled release [22].

7.1 METHOD OF CALCULATION

The mathematical treatment of the kinetics of the drug release from dosage forms controlled by erosion was developed in Chapter 4 (Section 4.2). Thus the equations have been established for various shapes of these dosage forms, expressing the kinetics of the drug release as it is determined through in vitro measurements.

7.1.1 ASSUMPTIONS FOR CALCULATING THE PLASMA DRUG PROFILE

- The process of drug transfer is illustrated in the Figure 7.1, where the following stages appear in succession: release of the drug from the dosage form along the GI tract, kinetics of absorption into the plasma compartment, and elimination from it.

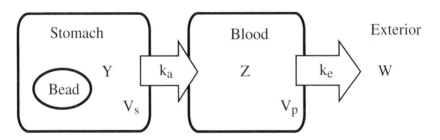

FIGURE 7.1 Scheme of the process of drug transfer from the dosage form into the plasma.

- The process of drug release from the dosage form is controlled by erosion, with a constant rate. (In fact, this strict assumption is not necessary provided that the process of the change in the rate of erosion along the GIT is known.)
- The phenomena of absorption and elimination are described by first-order kinetics with the two rate constants k_a and k_e expressed in h^{-1}.
- The rate constant of absorption remains constant along the GI tract time. (In fact, the numerical model can take into account any change in this rate constant.)
- The pharmacokinetics process is linear, meaning that the drug concentration in the plasma is proportional to the dose.

7.1.2 MATHEMATICAL AND NUMERICAL TREATMENT

From the kinetics shown in Chapter 4 (Section 4.2), the rate of drug delivery by the dosage form at time t, whatever the shape of the dosage form, is expressed by the derivative $\frac{dM}{dt}$.

The amount of drug in the GI is Y_t, which can be related to the rate constant of absorption k_a by the following equation:

$$\frac{dY}{dt} = \frac{dM}{dt} - k_a Y \qquad (7.1)$$

Then, the amount of drug in the plasma compartment Z_t is expressed by

$$\frac{dZ}{dt} = k_a Y - k_e Z \qquad (7.2)$$

and the amount of drug eliminated at time t is W:

$$\frac{dW}{dt} = k_e Z \qquad (7.3)$$

Because mathematical treatment is not feasible, the plasma drug level Z, as well as the kinetics of elimination, are evaluated by using a numerical method, run step by step, with a constant increment of time.

7.2 PLASMA DRUG PROFILE WITH A SINGLE DOSE

Various factors intervene in the case of oral dosage forms whose release is controlled by erosion: the shape of the dosage form and (much more important) the time of full release t_r. This time takes into account the dimensions of the dosage form and the nature of the drug with its pharmacokinetic parameters. These parameters are considered in succession.

7.2.1 EFFECT OF THE SHAPE OF THE DOSAGE FORMS

The dimensions of the dosage forms of various shapes are evaluated so that they have the same volume and the same amount of drug. A dosage form, spherical in shape, with a time of full erosion of 48 h, is taken as a reference (Figure 7.2, curve 4; Figure 7.3, curve 5). Other dosage forms of different shapes (i.e., parallelepiped, cube, and cylinder) are also considered for comparison.

The kinetics of drug release obtained with these four dosage forms through in vitro experiments is drawn in Figure 7.2 by plotting the amount of drug release at time t as a fraction of the amount initially in the dosage form as a function of time. The plasma drug profiles are calculated with these dosage forms (Figure 7.3).

Some relevant results are worth noting:

- The kinetics of drug release obtained with these dosage forms of various shapes (but having the same volume and the same amount of drug) are similar, with an oblique tangent at the origin of time, and the drug is totally released after a finite time (48 h).
- The shape given to the dosage forms plays a small role in the kinetics of release, this kinetics being faster with the parallelepiped and slower with the sphere.

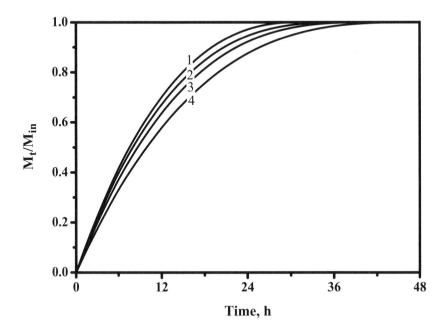

FIGURE 7.2 Kinetics of drug release controlled by erosion, as obtained through in vitro measurements, with various dosage forms of the same volume and different shapes. 1. Parallelepiped with $2a = 2b = c$. 2. Cube. 3. Cylinder with $R = H$. 4. Sphere, with $t_r = 48$ h for the sphere.

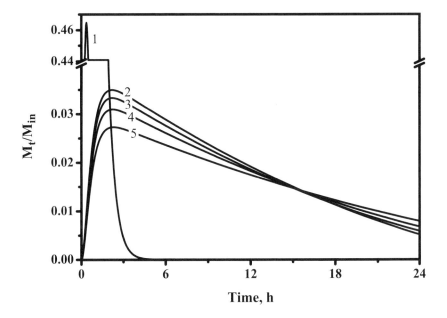

FIGURE 7.3 Plasma drug level ($\frac{M_t}{M_{in}}$) obtained with acetyl salicylic acid released from various dosage forms. 1. Immediate release. 2. Parallelepiped with $2a = 2b = c$. 3. Cube. 4. Cylinder with $R = H$. 5. Sphere, with $t_r = 48$ h for the sphere. $k_a = 3.5$/h and $k_e = 2.1$/h. A change in the ordinate is made for immediate release.

- The plasma drug levels obtained with erosion-controlled dosage forms are far more constant than the corresponding plasma drug profiles obtained with the immediate-release dosage form (Figure 7.3, curve 1) playing the role of reference.
- The effect of the shape of the dosage form on the plasma drug level can be appreciated by comparing the curves drawn in Figure 7.3. The most constant drug level is attained with the sphere (curve 5); this fact results from the longer erosion time and from the most linear outline of the kinetics of drug release (Figure 7.2, curve 4).
- By comparing the curves drawn in Figure 6.8 and Figure 7.3, when acetyl salicylic acid is the drug, it clearly appears that the plasma drug level is far more constant when the process of drug release is controlled by erosion than when it is controlled by diffusion [23]. The plasma drug profile can be extended to 24 h when the process is controlled by erosion, but it is not possible to obtain the same result with diffusion-controlled dosage forms.
- With a drug such as acetyl salicylic acid, whose rate constant of elimination is very high, the plasma drug concentration is very low.

7.2.2 Effect of the Time of Full Erosion

The effect of the time of full erosion on the plasma drug level is shown in Figure 7.4 (solid line) when acetyl salicylic acid is the drug, by considering the three values of

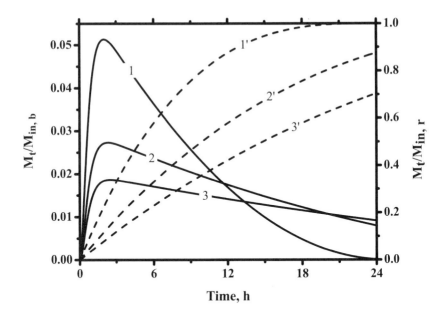

FIGURE 7.4 Plasma drug level ($\frac{M_L}{M_{in}}$) obtained with acetyl salicylic acid released from dosage forms, spherical in shape, having various values of the time of erosion t_r. 1. 24 h. 2. 48 h. 3. 72 h. (Solid lines), with $k_a = 3.5/h$ and $k_e = 2.1/h$. The kinetics of drug release from the corresponding dosage forms are noted 1', 2', 3' (dotted lines).

this time: 24, 48, and 72 h. In the same way, the kinetics of drug release from the dosage forms is also drawn (dotted lines).

The following conclusions can be traced from these curves:

- The time of full erosion plays a major role either in the kinetics of drug release or in the plasma drug profile. Of course, the lower value of the time of full erosion is responsible for a faster kinetics of release associated with the higher peak in the plasma drug profile. In contrast with this first case, a nearly constant plasma drug profile is obtained with longer times of full erosion, and especially with the time of 72 h (curve 3).
- Of course, an increase in the time of full erosion provokes also a decrease in the plasma drug concentration. This fact is especially true when the drug's rate constant of elimination is high, as in the case in hand with acetyl salicylic acid.
- It stands to reason that such dosage forms should be bioadhesive in order to maintain the dosage form in the GI over the period of time necessary for full erosion.

7.2.3 EFFECT OF THE NATURE OF THE DRUG

The effect of the nature of the drug on the plasma drug profile is studied by considering the time of full erosion of 48 h with the following drugs: ciprofloxacin (curve 1), cimetidine (curve 2), and acetyl salicylic acid (curve 3), whose pharma-cokinetics parameters are shown in the caption of Figure 7.5. These curves, express-ing the variation of the amount of drug in the plasma as a fraction of the amount initially in the dosage form with time, lead to a few comments:

- The plasma drug profile largely depends on the nature of the drug, and the statement holds: the larger the rate constant of elimination, the faster the rate of drug elimination and the lower the plasma drug concentration. This is especially obvious for these three drugs, whose rate constants of elimination are quite different.
- The peak for the profile of ciprofloxacin, which exhibits the lowest rate constant of elimination, is not only the largest, but it is also attained at the longest time.
- The rate constant of absorption does not intervene much in the plasma drug profile.

7.2.4 PATIENTS' INTERVARIABILITY

The effect of patients' intervariability is considered in evaluating the plasma drug profile by keeping the same time of full erosion of 48 h and by varying the value of the rate constant of elimination of the drugs (or the half-life time, which is

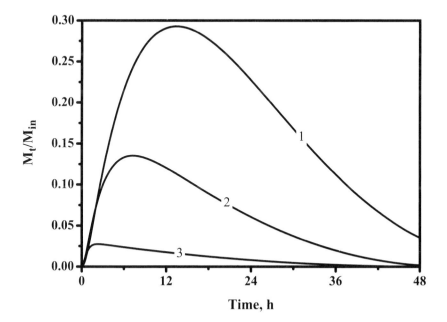

FIGURE 7.5 Plasma drug level $(\frac{M_t}{M_{in}})$ obtained with various drugs in spherical dosage forms and the same conditions. 1. Ciprofloxacin with $k_a = 1.3/h$ and $k_e = 0.12/h$. 2. Cimetidine with $k_a = 2.2/h$ and $k_e = 0.34/h$. 3. Acetyl salicylic acid with $k_a = 3.5/h$ and $k_e = 2.1/h$. Time of erosion $t_r = 48\,h$.

inversely proportional, as shown in Equation 7.4) within the range defined by various authors. Thus, the plasma drug profiles are drawn for ciprofloxacin in Figure 7.6 [24], for acetyl salicylic acid in Figure 7.7 [25], and for cimetidine in Figure 7.8 [26].

$$Ln2 = 0.693 = k_e t_{0.5} \qquad (7.4)$$

Some conclusions of great concern appear when it is desired to adapt the therapy to the patient:

- Primarily, the effect of the nature of the drug is of prime importance. Thus, the drug profile is extended over a period of time of 48 h for ciprofloxacin (Figure 7.6), 24 h only for acetyl salicylic acid (Figure 7.7), and 36 h for cimetidine (Figure 7.8). Of course, this fact results from the value of the rate constant of elimination, which is much larger for acetyl salicylic acid and much lower for ciprofloxacin.
- The effect of patients' intervariability toward the drugs results from the change in the rate constant of elimination. Thus, the statement holds: the larger the rate constant of elimination (or the shorter the half-life time), the lower the concentration of the drug in the plasma and the shorter the new half-life time obtained with the erosion-controlled dosage form. More precisely, these facts are written in Table 7.1 as they are measured from the curves drawn in Figure 7.6 through Figure 7.8. In the same way that

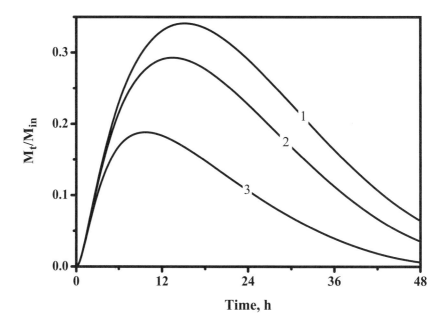

FIGURE 7.6 Plasma drug level ($\frac{M_t}{M_{in}}$) obtained with ciprofloxacin, in spherical dosage forms administered under the same conditions. Intervariability of the patients with $k_a = 1.3$/h and various values of k_e: 1. 0.09/h. 2. 0.12/h. 3. 0.22/h. $t_r = 48$ h.

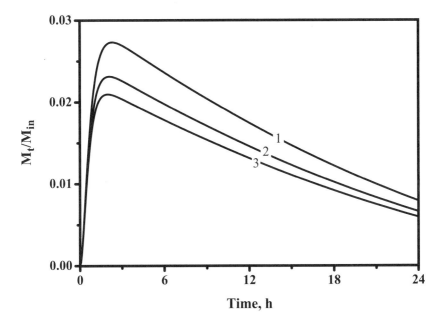

FIGURE 7.7 Plasma drug level ($\frac{M_t}{M_{in}}$) obtained with acetyl salicylic acid, in spherical dosage forms administered under the same conditions. Intervariability of the patients with $k_a = 3.5/h$ and various values of k_e: 1. 2.1/h. 2. 2.5/h. 3. 2.77/h. $t_r = 48$ h.

Table 6.3 was concerned with diffusion-controlled dosage forms, Table 7.1 shows that the half-life time obtained with these dosage forms, $T_{0.5}$, is the time for which the plasma drug concentration is half the value attained at the peak. The half-life time $T'_{0.5}$ is, in the same way, obtained with the immediate-release dosage form.

7.2.5 COMPARISON OF THE PROCESSES OF DIFFUSION AND EROSION

Comparison of the processes of drug release controlled either by diffusion or by erosion is of great interest. The maximum value at the peak defined by the ratio $\frac{M_{max}}{M_{in}}$, which is proportional to the maximum drug concentration, and the values of the half-life time $T_{0.5}$ obtained with controlled-release dosage forms can give a good idea of the release efficiency provided by these two release processes. These data are collected in Table 7.2, where they are evaluated with the parameter of diffusion $\frac{D}{R^2} = 5 \times 10^{-6}/s$ and with the time of erosion $t_r = 48$ h. The value for the parameter of diffusion allows the drug to be nearly totally released within 24 h, whereas the erosion time of 48 h means that the drug is totally released within 48 h.

The following conclusions are offered about these two processes of controlled drug release:

- Erosion time of 48 h can be selected only when the erosion-controlled dosage forms are made of bioadhesive polymers, meaning that these dosage

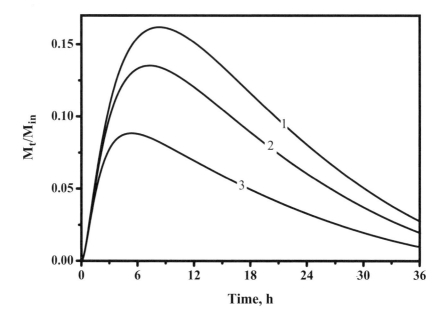

FIGURE 7.8 Plasma drug level ($\frac{M_t}{M_{in}}$) obtained with cimetidine, in spherical dosage forms administered under the same conditions. Intervariability of the patients with $k_a = 2.2$/h and various values of k_e: 1. 0.27/h. 2. 0.34/h. 3. 0.57/h. $t_r = 48$ h.

      forms can be kept in the GI over a period of time of 48 h, significantly exceeding the usual GIT.

- The value of the parameter of diffusion (e.g., $\frac{D}{R^2} = 5 \times 10^{-6}$/s) allows the dosage forms to release nearly all the drug over a time period of 24 h.
- From the preceding two considerations, it appears that there is an advantage for the erosion process when sustained release is desired.
- As shown in Table 7.2, the values of the half-life time are much larger with the process of erosion. This fact is especially sensible for drugs having a larger rate constant of elimination, such as acetyl salicylic acid.

TABLE 7.1
Pharmacokinetic Parameters of the Drugs

Drug	Ciprofloxacin [24]			Acetyl Salicylic Acid [25]			Cimetidine [26]		
k_a (/h)		1.3			3.5			2.2	
k_e (/h)	0.09	0.12	0.22	2.1	2.5	2.77	0.27	0.34	0.57
$t_{0.5}$ (h)	7.7	5.7	3.1	0.33	0.28	0.25	2.6	2	1.2
$T'_{0.5}$ (h)	10.5	8.4	5.7						
$T_{0.5}$ (h)	35.6	32.4	26.2	16.6	16.4	16.2	24	22.5	20.2
$\frac{M_{max}}{M_{in}}$	0.34	0.29	0.19	0.03	0.02	0.2	0.16	0.13	0.09

TABLE 7.2
Comparison between the Processes of Diffusion and Erosion

Drug	Ciprofloxacin [24]			Acetyl Salicylic Acid [25]			Cimetidine [26]		
k_a (/h)		1.3			3.5			2.2	
k_e (/h)	0.09	0.12	0.22	2.1	2.5	2.77	0.27	0.34	0.57
$t_{0.5}$ (h)	7.7	5.7	3.1	0.33	0.28	0.25	2.6	2	1.2

Process Controlled by Diffusion with $\frac{D}{R^2} = 5 \times 10^{-6}/s$

$T_{0.5}$ (h)	19.8	16.5	11.1	2.28	2.08	1.97	9.2	7.5	5.7
$\frac{M_{max}}{M_{in}}$	0.53	0.49	0.39	0.13	0.12	0.11	0.37	0.33	0.26
AUC	11.1	8.63	4.53	0.47	0.40	0.36	3.69	2.93	1.75

Process Controlled by Erosion with $t_r = 48$ h

$T_{0.5}$ (h)	35.6	32.4	26.2	16.6	16.4	16.2	24	22.5	20.2
$\frac{M_{max}}{M_{in}}$	0.34	0.29	0.19	0.03	0.02	0.2	0.16	0.13	0.09
AUC	11.1	8.65	4.54	0.47	0.40	0.36	3.70	2.94	1.75

However, the values of the drug concentration, as expressed by the ratio $\frac{M_t}{M_{in}}$, are obviously lower with the process of erosion.

- The AUC values are of interest. In fact, for each drug and with every value of the rate constant of elimination, the AUC values, expressed in terms of $\frac{M_t}{M_{in}}$ time (this time is in hours in Table 7.2), are similar: the difference appears only in the third digit. Recall that instead of the concentration of the drug, which makes mandatory the use of the apparent plasmatic volume, the amount of the drug in the plasma is expressed as a fraction of the amount of drug initially in the dosage form, $\frac{M_t}{M_{in}}$. Thus, the AUC values shown in Table 7.2 should be multiplied by the amount of drug initially in the dosage form and divided by the apparent plasmatic volume in order to be expressed as usual in terms of concentration–time.

7.3 PLASMA DRUG PROFILE WITH REPEATED MULTIDOSES

The following parameters should be examined in succession when the dosage forms are taken in repeated multidoses:

- The shape given to the dosage form
- The time of full erosion
- The nature of the drug
- The consequences resulting from patients' intervariability

7.3.1 EFFECT OF THE SHAPE OF THE DOSAGE FORM

The plasma drug levels, expressed in terms of the ratio $\frac{M_t}{M_{in}}$, obtained with dosage forms of same volume (and same amount of drug) and different shapes, whose release is controlled by erosion with the full erosion time, t_r of 48 h, are drawn in Figure 7.9. These profiles lead to some conclusions of interest:

- Obviously, the plasma drug levels obtained with erosion-controlled dosage forms are far more constant than the drug level obtained with the immediate-release dosage form.
- The effect of a dosage form's shape on the plasma drug level is not very significant. However, it can be said that the most constant drug level is attained with the sphere, this fact resulting from the longer time of erosion and from the shape of the kinetics of drug release, as shown in Figure 7.2, curve 4.
- It also appears clear that the plasma drug level is far more constant when the process of drug release is controlled by erosion than when it is controlled by diffusion. For instance, comparison of these two processes for the sphere can be made with the curve 2 in Figure 6.18 (Chapter 6), where the parameter of diffusion is $\frac{D}{R^2} = 5 \times 10^{-6}$/s, and with curve 5 in Figure 7.9, drawn with the parameter of erosion $t_r = 48$ h. The difference is so large between these profiles that the erosion-controlled dosage forms

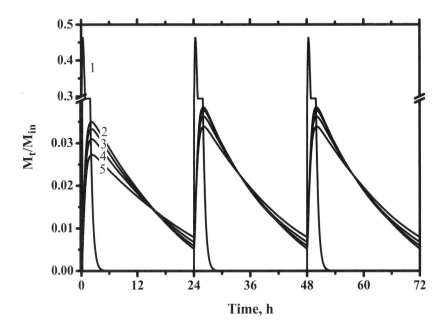

FIGURE 7.9 Plasma drug level ($\frac{M_t}{M_{in}}$) obtained with acetyl salicylic acid, with repeated doses taken once a day, and dosage forms of various shapes. 1. Immediate release. 2. Parallelepiped with $2a = 2b = c$. 3. Cube. 4. Cylinder with $R = H$. 5. Sphere with $t_r = 48$ h for the sphere. $k_a = 3.5$/h and $k_e = 2.1$/h. A change in the ordinate is made for immediate release.

can be taken once a day, whereas their counterparts controlled by diffusion should be taken four times a day in order to obtain similar profiles.

- Of course, as already said with single doses, the amount of drug at any time in the plasma obtained with erosion-controlled dosage forms is much lower than that obtained when the process is controlled by diffusion. Nevertheless, the *AUC* values are similar for the two processes, as shown in Table 7.2.

7.3.2 EFFECT OF THE TIME OF FULL EROSION

The effect of the time of full erosion t_r on the plasma drug level can be seen in Figure 7.10, with repeated doses taken once a day for different values of t_r: 24 h (curve 1), 48 h (curve 2), and 72 h (curve 3). Acetyl salicylic acid was selected for the drug, in dosage forms of same volume and spherical in shape.

The following conclusions can be made from these curves:

- The effect of the time of full erosion on the plasma drug profiles is of great concern, as already seen in Figure 7.4 with single doses and the same drug.
- This effect is so apparent that only dosage forms with a value of t_r larger than 24 h (e.g., 48 or 72 h) can be taken once a day without provoking too low a drug concentration at the trough.

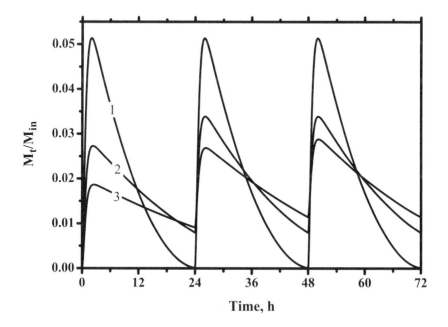

FIGURE 7.10. Plasma drug level ($\frac{M_t}{M_{in}}$) obtained with acetyl salicylic acid, with repeated doses taken once a day, with dosage forms spherical in shape, and various values of the time of erosion. 1. 24 h. 2. 48 h. 3. 72 h. $k_a = 3.5$/h and $k_e = 2.1$/h.

7.3.3 EFFECT OF THE NATURE OF THE DRUG

The effect of the nature of the drug clearly appears in Figure 7.11, which shows the plasma drug profiles obtained with dosage forms, spherical in shape, taken once a day, when the drug is ciprofloxacin (curve 1), cimetidine (curve 2), and acetyl salicylic acid (curve 3). These dosage forms are erosion controlled with a time of full erosion $t_r = 48$ h.

The conclusions are similar to those given for a single dose in Section 7.2.3 (Figure 7.5):

- The effect of the nature of the drug taken in repeated doses is so important that the profiles drawn in Figure 7.11 are quite different for these three drugs.
- For ciprofloxacin (curve 1), the dosage forms with $t_r = 48$ h, taken once a day, provoke an interesting plasma drug profile. This drug profile alternates between peaks and troughs that are not far apart, their ratio being as low as 1.56, whereas the peak concentration increases from dose to dose up to the third dose, at which the steady state is attained.
- For cimetidine (curve 2), with the same dosage form and $t_r = 48$ h, the steady state is reached at the second dose, and the ratio of the concentrations at peak and at trough is 2.7.

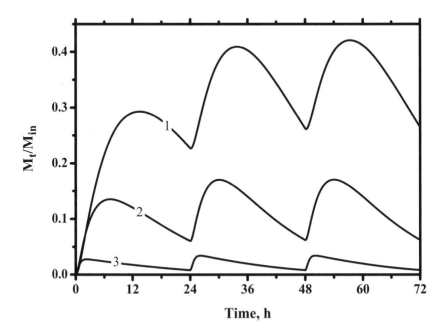

FIGURE 7.11 Plasma drug level ($\frac{M_t}{M_{in}}$) obtained with repeated doses taken once a day, with dosage forms spherical in shape with a time of erosion of 48 h, and various drugs. 1. Ciprofloxacin with $k_a = 1.3$/h and $k_e = 0.12$/h. 2. Cimetidine with $k_a = 2.2$/h and $k_e = 0.34$/h. 3. Acetyl salicylic acid with $k_a = 3.5$/h and $k_e = 2.1$/h.

- For acetyl salicylic acid (curve 3), the plasma concentration is very low, resulting from the very large rate constant of elimination, but the concentration at the troughs is not 0 when $t_r = 48$ h, because the ratio of the concentrations at peaks and at troughs does not exceed 3.6.

7.3.4 PATIENTS' INTERVARIABILITY

The effect of the intervariability of the patients is considered with the three drugs: ciprofloxacin (Figure 7.12), acetyl salicylic acid (Figure 7.13), and cimetidine (Figure 7.14), dispersed in erosion-controlled dosage forms having a time of full erosion $t_r = 48$ h, when they are taken once a day. The various values of the pharmacokinetic parameters are shown either in the captions of these three figures or in Table 7.1.

The results are similar to those obtained with a single dose, with the additional effect resulting from the repeated doses:

- With ciprofloxacin, the effect of patients' intervariability is of great concern, as shown in Figure 7.12. For the lower value of the rate constant of elimination k_e (curve 1), the ratio of the peak to the troughs tends to 1.46 at the steady state, which is attained at the third dose. With the larger value of k_e (curve 3), the same peak-to-trough ratio is 2.2 at the steady state, which is reached at the second dose. But the most important effect of patients'

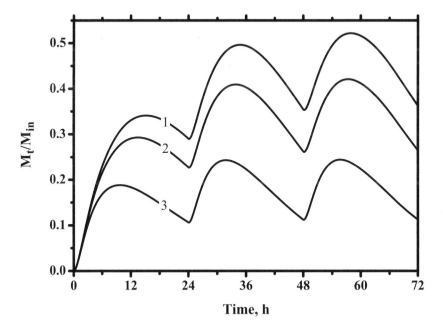

FIGURE 7.12 Plasma drug level ($\frac{M_L}{M_{in}}$) obtained with repeated doses taken once a day, with dosage forms spherical in shape with a time of erosion of 48 h, and ciprofloxacin. Effect of patients' intervariability with $k_a = 1.3$/h and various values of k_e. 1. 0.09/h. 2. 0.12/h. 3. 0.22/h.

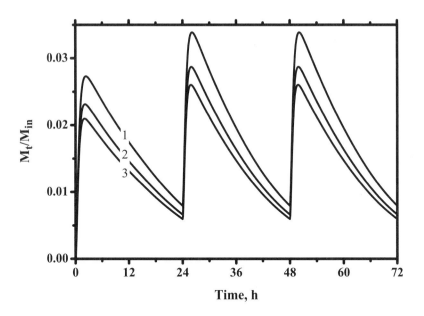

FIGURE 7.13 Plasma drug level ($\frac{M_t}{M_{in}}$) obtained with repeated doses taken once a day, with dosage forms spherical in shape with a time of erosion of 48 h, and acetyl salicylic acid. Effect of patients' intervariability with $k_a = 3.5$/h and various values of k_e. 1. 2.1/h. 2. 2.5/h. 3. 2.77/h.

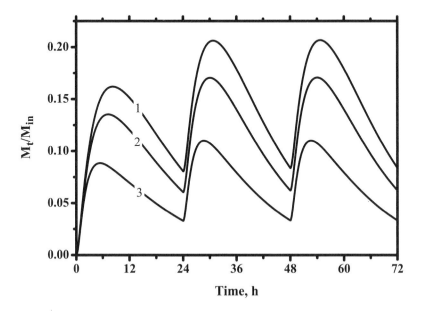

FIGURE 7.14 Plasma drug level ($\frac{M_t}{M_{in}}$) obtained with repeated doses taken once a day, with dosage forms spherical in shape with a time of erosion of 48 h, and cimetidine. Effect of patients' intervariability with $k_a = 2.2$/h and various values of k_e. 1. 0.27/h. 2. 0.34/h. 3. 0.57/h.

intervariability stands in the ratio of the plasma drug concentrations obtained with these two extreme categories of patients (curves 1 and 3): the ratio of the peaks is 2.1, and the ratio of the troughs is as high as 3.2.

- For acetyl salicylic acid (Figure 7.13), the effect of patients' intervariability is not as important as it is for ciprofloxacin, because the extreme values of the rate constant of elimination do not diverge so much. Thus, the steady state for all the patients is attained from the second dose, and the plasma drug levels are in the ratio of 1.3.

- For cimetidine (Figure 7.14), the change in the plasma drug profiles is very important, resulting from the large variation between the extreme values of the rate constant of elimination. The steady state is attained from the second dose in all cases, and the peak–trough ratio is 2.5 for the lower value of k_e and 3.4 for the larger value.

7.4 PREDICTION OF THE CHARACTERISTICS OF THE DOSAGE FORMS

It is of interest to predict by calculation the right characteristics of the dosage forms that can deliver the drug with the desired therapy. Section 7.4.1 and Section 7.4.2 are developed with two objectives: correlation between the half-life times and the plasma drug profiles expressed in terms of drug concentration.

7.4.1 Relationships between the Half-Life Times $t_{0.5}$ and $T_{0.5}$

Based on the process of erosion, the importance of various parameters, such as the nature of the drug, patients' intervariability (which corresponds to the response of the patient to the drug in terms of pharmacokinetics), and the shape and rate of erosion of the erodible polymer, has been explored. Thus the effect of the time of full erosion t_r has been defined, not only with single doses, but also in repeated doses, with the important factor of dose frequency.

In the same way as for diffusion-controlled release systems, relationships are built so as to connect the half-life time of the drug $t_{0.5}$ measured through i.v. delivery with the half-life time of the drug $T_{0.5}$ obtained when dispersed in the erodible polymer of the dosage form.

Two diagrams show the characteristics of the dosage form that are necessary to obtain the desired value of the half-life time of the drug $T_{0.5}$ dispersed in the erodible polymer, as a function of the half-life time of the drug itself $t_{0.5}$ measured through i.v. Recall that the time of full erosion t_r is defined by Equation 7.5:

$$t_r = \frac{R}{v} = \frac{L}{v} \tag{7.5}$$

where R is the radius of the dosage form, spherical in shape, and L is the smaller dimension of dosage forms having another shape.

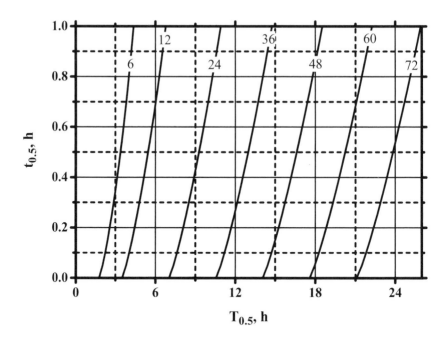

FIGURE 7.15 Diagram showing the relation between the half-life times $t_{0.5}$ and $T_{0.5}$ for various values of the time of erosion of dosage forms, spherical in shape, when $0 < t_{0.5} < 1$.

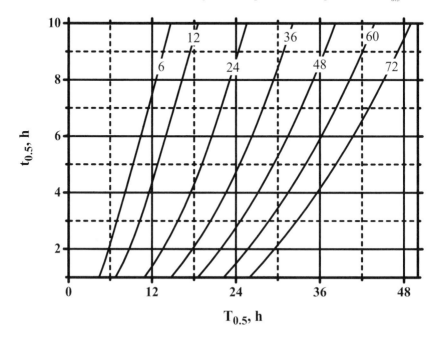

FIGURE 7.16 Diagram showing the relation between the half-life times $t_{0.5}$ and $T_{0.5}$ for various values of the time of erosion of dosage forms, spherical in shape, when $1 < t_{0.5} < 10$.

Thus, for a given rate of erosion v, the main dimension of the dosage form can be adjusted using Equation 7.5.

The first diagram in Figure 7.15 is drawn for drugs with a low half-life time $t_{0.5}$, such as acetyl salicylic acid. The second diagram in Figure 7.16 is concerned with drugs having much larger half-life times.

From these diagrams, it can be seen that the drug plays a major role, even dispersed through a polymer, and that it is not easy to prepare dosage forms capable of delivering the drug over a long period of time when this drug has a very short half-life time, as was already shown for a single dose in Figure 7.4 (Section 7.2.2) and for repeated doses in Figure 7.10 (Section 7.3.2).

7.4.2 PLASMA DRUG PROFILES IN TERMS OF CONCENTRATION

As previously stated in the case of diffusion-controlled dosage forms, it is easy to transform by reckoning the ordinate $\frac{M_t}{M_{in}}$ in terms of plasma drug concentration, by introducing the amount of drug M_{in} initially in the dosage form and dividing this amount M_t by the apparent plasmatic volume. The plasma drug profiles are thus drawn by considering the plasma drug concentration, with ciprofloxacin dispersed in dosage forms, spherical in shape, with a time of full erosion of either 48 h (Figure 7.17) or 24 h (Figure 7.18), when the amount of the drug is 500 mg.

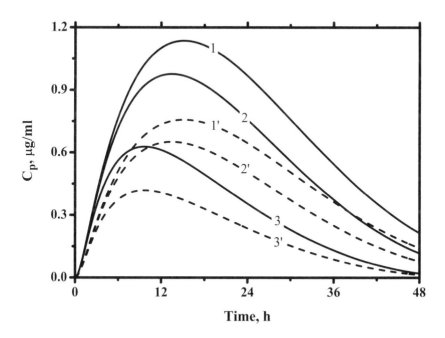

FIGURE 7.17 Plasma drug level concentration C_t obtained with repeated doses taken once a day, with dosage forms, spherical in shape, having a time of erosion of 48 h, and containing 500 mg ciprofloxacin. Effect of patients' intervariability with $k_a = 1.3$/h and various values of k_e and of the apparent plasmatic volume: (1 and 1') 0.09/h; (2 and 2') 0.12/h; (3 and 3') 0.22/h. (1, 2, 3) 150 l; (1', 2', 3') 225 l.

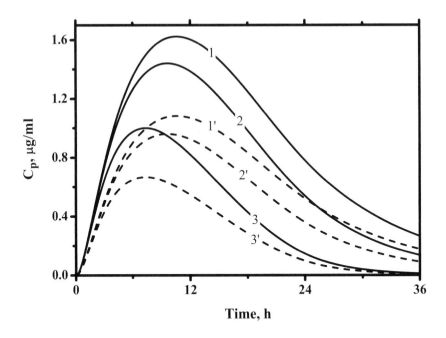

FIGURE 7.18 Plasma drug level concentration C_t obtained with repeated doses taken once a day, with dosage forms, spherical in shape, having a time of erosion of 24 h, and containing 500 mg ciprofloxacin. Effect of patients' inter-variability with $k_a = 1.3$/h and various values of k_e and of the apparent plasmatic volume: (1 and 1') 0.09/h; (2 and 2') 0.12/h; (3 and 3') 0.22/h. (1, 2, 3) 150 l; (1', 2', 3') 225 l.

The following conclusions can be drawn from these curves:

- Of course, the curves drawn in Figure 7.17 calculated with the time of full erosion of 48 h are similar to those drawn in Figure 7.6 obtained under the same conditions and expressed with the ordinate $\frac{M_t}{M_{in}}$.
- The effect of the time of full erosion clearly appears by comparing the curves in Figure 7.17 and Figure 7.18. In abscissa, the time over which the drug is delivered is 48 h when the time of full erosion is 48 h, and only 36 h when t_r is 24 h.
- Patients' intervariability is also considered, with three values of the rate constant of elimination (curves 1; 2; 3) and two values of the apparent plasmatic volume (solid lines with 150 l; dotted lines with 225 l).

REFERENCES

1. Vergnaud, J.M., *Controlled Drug Release of Oral Dosage Forms,* Ellis Horwood, London, 1993, Chapter 12.
2. Heller, J., Biodegradable polymers in controlled drug delivery, *CRC Crit. Rev. Therap. Drug Carrier Syst.,* 1, 39, 1984.

3. Vergnaud, J.M., *Liquid Transport Processes in Polymeric Materials,* Prentice Hall, Englewood Cliffs, NJ, 1991, Chapter 10.
4. Bidah, D. and Vergnaud, J.M., Kinetics of in-vitro release of Na salicylate dispersed in Gelucire, *Int. J. Pharmacol.,* 58, 215, 1990.
5. Bonferoni, M.C. et al., Influence of medium on dissolution-erosion behaviour of Na carboxymethylcellulose and on viscoelastic properties of gels, *Int. J. Pharm.,* 117, 41, 1995.
6. Falson-Rieg, F., Faivre, V., and Pirot, F., *Nouvelles formes médicamenteuses,* Tec & Doc Publ., Paris, 2004, Chapter 1.
7. Peppas, N.A. and Sahlin, J.J., Hydrogels as mucoadhesive and bioadhesive materials: a review, *Biomaterials,* 17, 1553, 1996.
8. Ponchel, G. et al., Mucoadhesion of colloidal particulate systems in the gastro-intestinal tract, *Eur. J. Pharm. Biopharm.,* 44, 25, 1997.
9. Sakuma, S. et al., Mucoadhesion of polystyrene nanoparticles having surface hydrophilic polymeric chains in the gastro-intestinal tract, *Int. J. Pharm.,* 177, 161, 1999.
10. Lavelle, E.C., Targeted delivery of drugs to the gastro-intestinal tract, *Crit. Rev. Ther. Drug,* 18, 341, 2001.
11. Lehr, C.M., From sticky stuff to sweet receptors: achievements, limits and novel approaches to bioadhesion, *Eur. J. Drug Metab. Pharmacokinet.,* 21, 139, 1996.
12. Kocevar-Nared, J., Kristl, J., and Smid-Korbar, J., Comparative rheological investigation of crude gastric mucin and natural gastric mucus, *Biomaterials,* 18, 677, 1997.
13. Madsen, F., Eberth, K., and Smart, J.D., A rheological examination of the mucoadhesive/mucus interaction: the effect of mucoadhesive type and concentration, *J. Control. Release,* 50, 167, 1998.
14. Riley, R.G. et al., An investigation of mucus/polymer rheological synergism using synthesised and characterised poly(acrylic acid)s, *Int. J. Pharm.,* 217, 87, 2001.
15. Jabbari, E. and Peppas, N.A., Molecular weight and polydispersity effects at the interface between polystyrene and poly(vinyl methyl ether), *J. Mater. Sci.,* 29, 3969, 1994.
16. Repka, M.A. and McGinity, J.W., Physical–mechanical, moisture absorption and bioadhesive properties of hydroxy propyl cellulose hot-melt extruded films, *Biomaterials,* 21, 1509, 2000.
17. Rossi, S. et al., Model-based interpretation of creep profiles for the assessment of polymer–mucin interaction, *Pharm. Res.,* 16, 1456, 1999.
18. Marshall, P. et al., A novel application of NMR microscopy: measurement of water diffusion inside bioadhesive bonds, *Magn. Resonance Imaging,* 19, 487, 2001.
19. Kockisch, S. et al., A direct staining method to evaluate the mucoadhesion of polymers from aqueous dispersion, *J. Control. Release,* 77, 1, 2001.
20. Quintanar-Guerrero, D. et al., In vitro evaluation of the bioadhesive properties of hydrophobic polybasic gels containing N-dimethylaminoethyl methacrylate-co-methylmethacrylate, *Biomaterials,* 22, 957, 2001.
21. Riley, R.G. et al., An in vitro model for investigating the gastric mucosal retention of C^{14}-labelled poly(acrylic acid) dispersions, *Int. J. Pharm.,* 236, 87, 2002.
22. Duchêne, D. and Ponchel, G., Bioadhesion of solid oral dosage forms: why and how? *Eur. J. Pharm. Biopharm.,* 44, 15, 1997.
23. Aïnaoui, A. and Vergnaud, J.M., Effect of the nature of the polymer and of the process of drug release (diffusion or erosion) for oral dosage forms, *Computat. Theoret. Polymer Sci.,* 10, 383, 2000.

24. Fabre, D. et al., Steady-state pharmacokinetics of ciprofloxacin in plasma from patients with nosocomial pneumonia: penetration of the bronchial mucosa, *Antimicrob. Agents Chemother.,* Dec. 1991, 2521, 1991.
25. Rowland, M. and Riegelman, S., Pharmacokinetics of acetylsalicylic acid and salicylic acid after intravenous administration in man, *J. Pharm. Sci.,* 57, 1313, 1968.
26. Somogyi, A. and Roland, G., Clinical pharmacokinetics of cimetidine, *Clin. Pharmacokin.,* 8, 463, 1983.

8 Effect of the Patient's Noncompliance

NOMENCLATURE

C_t Plasma drug concentration at time t.
D Diffusivity of the drug through the diffusion-controlled release dosage form.
$\frac{D}{R^2}$ Parameter of the dosage form whose drug release is controlled by diffusion (per s).
k_a Rate constant of absorption of the drug, expressed in h^{-1}.
k_e Rate constant of elimination of the drug, expressed in h^{-1}.
M_t Amount of drug delivered in the plasma up to time t.
M_{in} Amount of drug initially in the dosage form.
$\frac{M_t}{M_{in}}$ Dimensionless number, expressing the amount of drug delivered in the plasma up to time t.
t_r Time of full erosion of dosage forms whose release is controlled by erosion.
Acetyl salicylic acid $k_a = 3.5/h$ $k_e = 2.1/h$.
Ciprofloxacin $k_a = 1.3/h$ $k_e = 0.12/h$.
Cimetidine $k_a = 2.2/h$ $k_e = 0.34/h$.
V_p Apparent plasmatic volume.

8.1 LIMITED RELIABILITY OF THE PATIENT

Generally, physicians are unaware of how frequently their patients, unconsciously or deliberately, alter the dosage schedule prescribed. Various studies have clearly demonstrated that physicians hold overly optimistic views concerning this undistinguished problem. Some precise investigations conducted in the United States prove that not less than 40 to 60% of patients fail to comply with the prescribed regimen [1, 2]. To put it succinctly, patient compliance is worse than many physicians believe, dependent on the disease pattern, and variable with respect to the treating physician. More precisely, it was found that this percentage varies from disease to disease, and there is no clear-cut correlation between noncompliance and individual factors such as education and social status or gender and age.

Six factors intervene in noncompliance:

- Omission
- Wrong dosage
- Wrong frequency
- Impermissible delay
- Execution in the wrong mode
- Execution at the wrong time

The main causes of noncompliance refer primarily to the following three factors:

- Nature of the disease
- Type of medication
- Interaction of the physician or pharmacist with the patient

8.1.1 NATURE OF THE DISEASE

The character of the disease significantly influences patient behavior. Severe symptoms and obvious signs of the relentless progress of a cruel disease cause a consciousness of illness that exerts psychological pressure on the patient. This pressure is able to stimulate great reliability in following the prescribed treatment. On the other hand, patients have more difficulty in complying with the treatment of a disease that causes no distinct complaint, such as high blood pressure. Moreover, it seems difficult for patients to appreciate the necessity of the therapy when this therapy causes disturbing side effects, while the disease itself gives rise to little complaint.

8.1.2 TYPE OF MEDICATION

The facility of the patient's cooperation with the physician decreases regularly with the duration of the treatment. Willingness to cooperate also decreases with increasing complexity of the regimen, length of the treatment, and complexity of dosing prescriptions. These problems become particularly serious in chronic diseases.

8.1.3 INFLUENCES ON THE PATIENTS

It has been found that the influence of the physician and the pharmacist is decisive in a patient's compliance [2]. The patient must be informed precisely of the following:

- Why the drug must be taken
- How often
- How long
- At what dose

The treating physician is the person responsible for providing information on the why and how of the drug administration; the pharmacist shares this responsibility in adding his or her professional support.

8.1.4 VARIOUS CASES OF NONCOMPLIANCE

Forgetfulness, the first case of noncompliance, can have serious consequences in drug treatment. It should be said that if the patient omits one or two doses, this omission cannot be repaired by any subsequent regular dosage or, worse, by an adapted dosage determined by simple calculation.

Generally, patients do not admit their noncompliance. The problem of limited reliability has been studied only very recently. But data obtained from these studies make it necessary to reach realistic conclusions and act upon them. Therapeutic systems certainly offer the opportunity to improve patient compliance by reducing some of the difficulties connected with conventional dosage forms. First, all compliance studies make it clear that reducing the frequency of drug administration improves compliance, once-a-day dosing being the ideal objective. Second, the patient would understand that increasing compliance is associated with reducing bad and painful side effects.

Moreover, a simple way to convince patients that compliance is beneficial and that noncompliance may be dangerous is to show them the resulting plasma drug levels obtained with omission or the wrong dosage at the wrong time. This is especially obvious and necessary with sustained-release dosage forms containing a large amount of drugs.

As far as we know, only a few quantitative studies have been done on the problem of noncompliance by considering the process of drug delivery, in order to evaluate the change in the plasma drug profiles resulting from patients' noncompliance [3].

8.2 NONCOMPLIANCE WITH EROSION-CONTROLLED DOSAGE FORMS

8.2.1 PERFECT COMPLIANCE AND OMISSION

Perfect compliance with once-a-day dosing is associated with regular plasma drug profiles, as depicted in Figure 8.1. Calculation is made with the following three

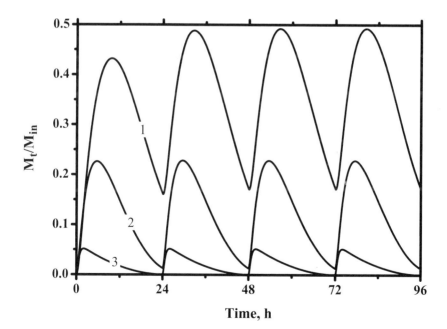

FIGURE 8.1 Plasma drug profiles, expressed in terms of the dimensionless number $\frac{M_t}{M_{in}}$, with perfect compliance with three drugs taken once a day, with erosion-controlled dosage forms with a time of erosion of 24 h. 1. Ciprofloxacin. 2. Cimetidine. 3. Acetyl salicylic acid.

drugs, whose pharmacokinetic parameters are shown in the Nomenclature:

1. Ciprofloxacin
2. Cimetidine
3. Acetyl salicylic acid

The dosage forms are made of the same polymer, whose time of erosion is 24 h, and the plasma drug profiles are expressed in terms of time by using the dimensionless number $\frac{M_t}{M_{in}}$ for the ordinate (e.g., the amount of drug in the plasma at time t as a fraction of the amount of drug initially in the dosage form). In Figure 8.1, the plasma drug profiles alternate between peaks and troughs at rather high levels for ciprofloxacin, with a peak–trough ratio of 2.9, while the same ratio is much higher, at not less than 17 for cimetidine. For acetyl salicylic acid, the dose frequency is not adapted, because of the high value of its rate of elimination, as shown in Chapter 7.

The effect of omission appears in Figure 8.2, with the same drugs dispersed in the same dosage forms with a time of erosion, t_r of 24 h, as those used in Figure 8.1. The omission taking place on the second day provokes a deep trough for all drugs, even for ciprofloxacin, which has the lowest rate of elimination, and on the third day, the right dosing again starts providing a plasma drug profile as normal as in Figure 8.1.

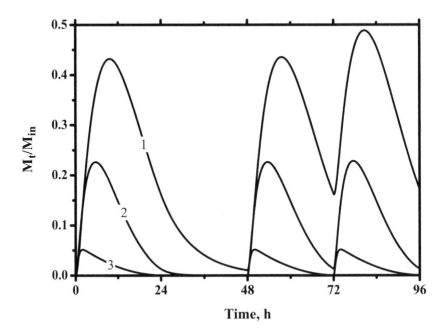

FIGURE 8.2 Plasma drug profiles, expressed in terms of the dimensionless number $\frac{M_t}{M_{in}}$, with noncompliance resulting from omission on the second day, with three drugs taken once a day, with erosion-controlled dosage forms with a time of erosion of 24 h. 1. Ciprofloxacin. 2. Cimetidine. 3. Acetyl salicylic acid.

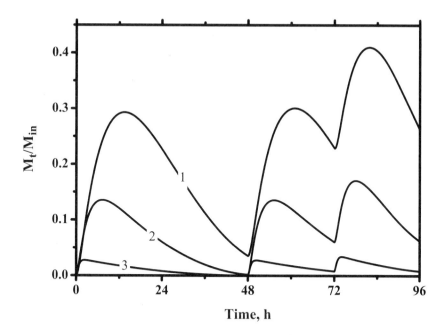

FIGURE 8.3 Plasma drug profiles, expressed in terms of the dimensionless number $\frac{M_t}{M_{in}}$, with noncompliance resulting from omission on the second day, with three drugs taken once a day, with erosion-controlled dosage forms with a time of erosion of 48 h. 1. Ciprofloxacin. 2. Cimetidine. 3. Acetyl salicylic acid.

In Figure 8.3, with a time of erosion of 48 h, and Figure 8.4, with a time of erosion of 72 h, respectively, the effect of the time of erosion clearly arises, leading to the following conclusions:

- The longer the erosion time of the polymer through which the drug is dispersed, the lower the effect of omission. In fact, with ciprofloxacin (curve 1), the trough resulting from the omission on the second day remains high, and it is not so low for cimetidine (curve 2).
- The larger the rate constant of elimination of the drug, the lower the trough resulting from the omission, as shown in Chapter 7. Thus, for acetyl salicylic acid (curve 3), there is a small amount of drug in the plasma over a period of nearly one day.

8.2.2 OMISSION FOLLOWED BY A DOUBLE DOSE

Following an omission, some patients, perhaps interested in arithmetic calculation, believe in compensating for this omission in taking a double dose afterwards. As a matter of fact, rather than compensating for the error, this is a wrong dosage in addition to the previous omission. According to Benjamin Franklin, "Lost time is never found again," and this proverb, well known in various countries, is also of value in therapy.

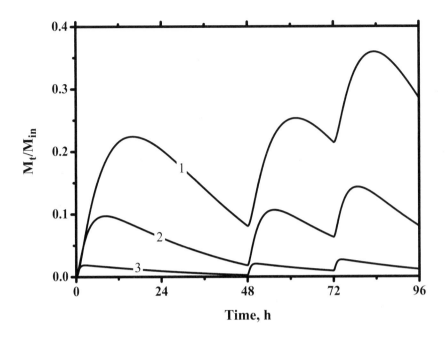

FIGURE 8.4 Plasma drug profiles, expressed in terms of the dimensionless number $\frac{M_t}{M_{in}}$, with noncompliance resulting from omission on the second day, with three drugs taken once a day, with erosion-controlled dosage forms with a time of erosion of 72 h. 1. Ciprofloxacin. 2. Cimetidine. 3. Acetyl salicylic acid.

This problem is studied by modeling the process, leading to quantitative results. The plasma drug profiles obtained in the case of omission on the second day and attempted compensation in taking a double dose on the third day are drawn for the three drugs, when they are orally delivered through sustained-release dosage forms:

1. Ciprofloxacin
2. Cimetidine
3. Acetyl salicylic acid

The process of release is controlled by erosion with three different values for the time of erosion: 24 h (Figure 8.5), 48 h (Figure 8.6), and 72 h (Figure 8.7). The plasma drug profiles are expressed as a function of time, by using the dimensionless number in ordinate, which is the amount of drug released up to time t as a fraction of the amount of drug initially in the dosage form. Moreover, the plasma drug concentration is also evaluated and drawn in Figure 8.8 for acetyl salicylic acid, with an erosion time of 72 h, the apparent plasma volume being 10 l and the quantity of drug of 1 g (curve 1) and 2 g (curve 2).

The following comments can be made based on Figure 8.5 through Figure 8.8:

- From the first approach, it clearly appears that after a lack of drug in the plasma resulting from the omission on the second day, there is an excess

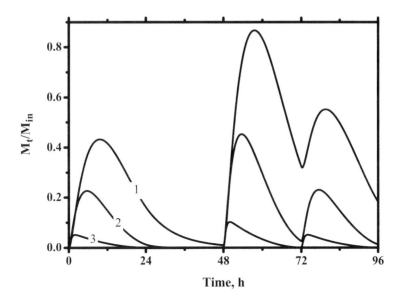

FIGURE 8.5 Plasma drug profiles, expressed in terms of the dimensionless number $\frac{M_t}{M_{in}}$, with noncompliance resulting from omission on the second day and double dose on the third day, with three drugs taken once a day, with erosion-controlled dosage forms with a time of erosion of 24 h. 1. Ciprofloxacin. 2. Cimetidine. 3. Acetyl salicylic acid.

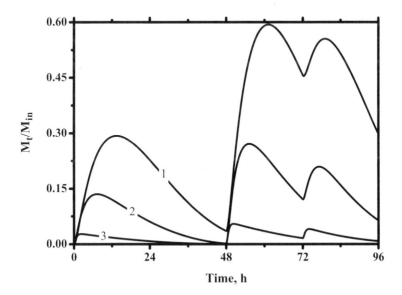

FIGURE 8.6 Plasma drug profiles, expressed in terms of the dimensionless number $\frac{M_t}{M_{in}}$, with noncompliance resulting from omission on the second day and double dose on the third day, with three drugs taken once a day, with erosion-controlled dosage forms with a time of erosion of 48 h. 1. Ciprofloxacin. 2. Cimetidine. 3. Acetyl salicylic acid.

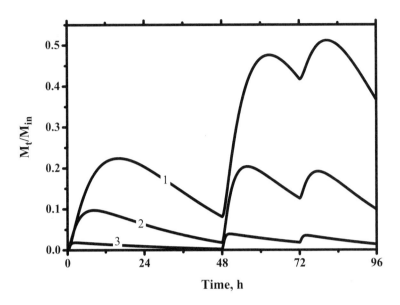

FIGURE 8.7 Plasma drug profiles, expressed in terms of the dimensionless number $\frac{M_t}{M_{in}}$, with noncompliance resulting from omission on the second day and double dose on the third day, with three drugs taken once a day, with erosion-controlled dosage forms with a time of erosion of 72 h. 1. Ciprofloxacin. 2. Cimetidine. 3. Acetyl salicylic acid.

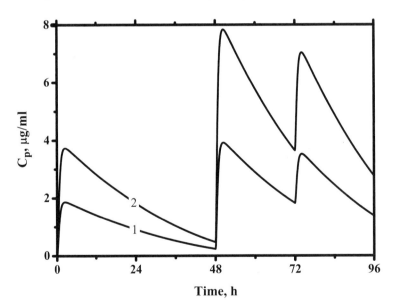

FIGURE 8.8 Plasma drug profiles, expressed in terms of concentration, with noncompliance resulting from omission on the second day and double dose on the third day, with acetyl salicylic acid as the drug in erosion-controlled dosage forms with a time of erosion of 72 h, with the apparent plasmatic volume of 10 l, and dose of 1 g (curve 1) and of 2 g (curve 2).

of drug on the third day, resulting from double dosing. On the whole, the peak on the third day could be expected to be twice as high as the peaks obtained under normal, correct conditions.

- The time of full erosion t_r is found to play an important role on the process. In fact, the statement holds: the longer the time of erosion, the more important the effect of the wrong dosage. For instance, with $t_r = 24$ h, the fourth peak is nearly as high as the first one. For a larger value of this time of erosion, such as 48 h, the fourth peak itself increases in height and becomes as high as the third peak associated with the double dose. Finally, when $t_r = 72$ h, the fourth peak is increased so much that it is higher than the third peak associated with the double dose.

- The nature of the drug, with its rate constant of elimination, plays a role on the effect of omission and double dosing. It could be said that the higher the rate of elimination of the drug, the lower the trough resulting from omission; moreover, the lower the rate constant of elimination of the drug, the larger the peak associated with double dosing.

- Thus, there is a twofold effect of the erosion time of the polymer, because this parameter plays an important role on the plasma drug profile not only in the case of omission but also in the wrong dosage when the double dose is taken in an attempt to compensate for the omission. In omission, a longer time of erosion is beneficial to the plasma drug level, as well as a lower rate constant of elimination of the drug, in the sense that it allows a more regular plasma drug profile. On the other hand, in double dosing, a longer time of erosion, as well as a lower rate constant of elimination of the drug, disturbs the plasma drug profile over a longer period of time.

8.3 NONCOMPLIANCE WITH DIFFUSION-CONTROLLED DOSAGE FORMS

8.3.1 RIGHT COMPLIANCE AND OMISSION

Just as with erosion-sustained release dosage forms, the processes of right compliance and omission are considered with oral dosage forms whose release is controlled by diffusion. The plasma drug profiles with the same drugs are drawn in Figure 8.9 when the dosage forms are taken regularly, when the diffusion parameter of the dosage form and polymer is $\frac{D}{R^2} = 10^{-6}$/s:

1. Once a day for ciprofloxacin
2. Twice a day for cimetidine
3. Four times a day for acetyl salicylic acid.

These profiles are also drawn in the case of omission on the second day in Figure 8.10. Comparison of the curves in Figure 8.9 and Figure 8.10 leads to conclusions of interest:

- As already seen from Chapter 6 and Chapter 7, the process of diffusion is less efficient than the process of erosion in the matter of sustained

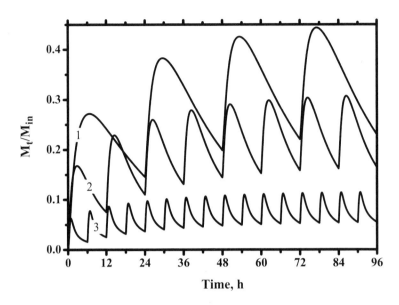

FIGURE 8.9 Plasma drug profiles, expressed in terms of the dimensionless number $\frac{M_t}{M_{in}}$, with perfect compliance with three drugs taken as follows: 1. once a day for ciprofloxacin; 2. twice a day for cimetidine; and 3. four times a day for acetyl salicylic acid, with diffusion-controlled dosage forms with $\frac{D}{R^2} = 10^{-6}$/s.

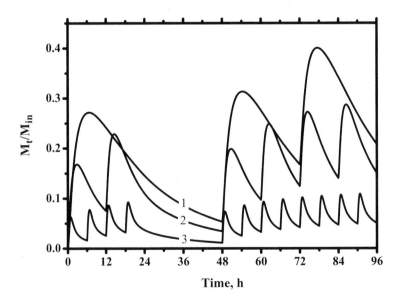

FIGURE 8.10 Plasma drug profiles, expressed in terms of the dimensionless number $\frac{M_t}{M_{in}}$, with noncompliance resulting from omission on the second day, with three drugs taken as follows: 1. once a day for ciprofloxacin; 2. twice a day for cimetidine; and 3. four times a day for acetyl salicylic acid, with diffusion-controlled dosage forms with $\frac{D}{R^2} = 10^{-6}$/s.

release. Thus in order to obtain flat plasma drug profiles, it is necessary to take cimetidine twice a day and acetyl salicylic acid four times a day. When the compliance is perfect, the steady state is attained after the fifth dose under these conditions of dosing.

- Under these dosing conditions, the effect of omission in not dramatic; the plasma drug level falls to a value that is not so low, being around 20% of the value attained at the trough under steady state without omission, for all three drugs.
- Of course, after the omission on the second day, from the third day when the doses are taken regularly, the plasma drug level increases continuously up to the steady state in a way similar to that shown in Figure 8.9.

8.3.2   OMISSION FOLLOWED BY DOUBLE DOSING

The process described in Section 8.3.1 is considered with the following change: the dosage forms are taken the first day, with an omission the second day, and all the doses taken on the third day are doubled, leading to the profiles drawn in Figure 8.11 for the three drugs. Another way of double dosing is also studied in the case of

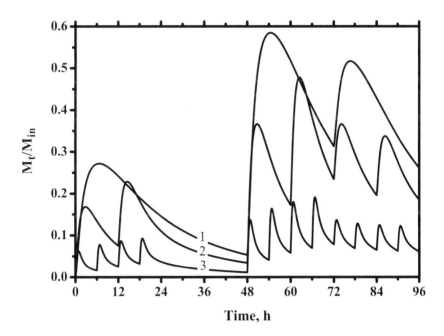

FIGURE 8.11 Plasma drug profiles, expressed in terms of the dimensionless number $\frac{M_t}{M_{in}}$, with noncompliance resulting from omission on the second day and doubling each dose on the third day, with three drugs taken as follows: 1. once a day for ciprofloxacin; 2. twice a day for cimetidine; and 3. four times a day for acetyl salicylic acid, with diffusion-controlled dosage forms with $\frac{D}{R^2} = 10^{-6}$/s.

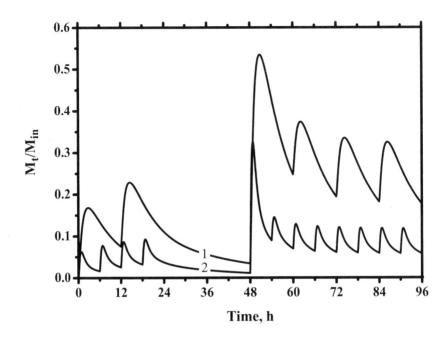

FIGURE 8.12 Plasma drug profiles, expressed in terms of the dimensionless number $\frac{M_t}{M_{in}}$, with noncompliance resulting from omission on the second day, and increasing as follows: 1. threefold the first dose on the third day, with cimetidine; 2. increasing five times the first dose on the third day for acetyl salicylic acid, with diffusion-controlled dosage forms with $\frac{D}{R^2} = 10^{-6}/s$, while cimetidine is taken twice a day and acetyl salicylic acid is taken four times a day.

cimetidine and acetyl salicylic acid, which are taken more frequently than once a day (Figure 8.12), since the first dose taken on the third day is 3 times the original dose for cimetidine and five times for acetyl salicylic acid, while the following doses are normal. In this worst case of wrong dosing, the patient takes, in the first dose on the third day, all the drug that was omitted on the second day. From the plasma drug profiles drawn in Figure 8.11 through Figure 8.13, a few remarks are worth making:

- Of course, the approach of taking the dose on the third day in order to compensate for the amount of drug omitted on the second day is worse when the total amount of the drug omitted is taken in the first dose, as shown in Figure 8.12 for cimetidine (curve 1) and acetyl salicylic acid (curve 2).
- The effect of double dosing on the third day is quite different from one drug to another. This difference is due essentially to the frequency of dosing, which has been selected according to the nature of the drug.
- The amount of drug at the peak resulting from the double dosing on the third day (Figure 8.12, curve 1) is around 1.8 times larger than the corresponding value under normal steady state conditions shown in Figure 8.9, curve 3.

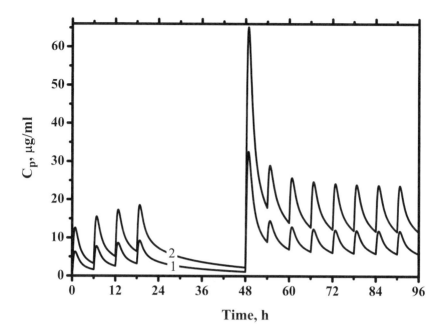

FIGURE 8.13 Plasma drug profiles, expressed in terms of concentration, with noncompliance resulting from omission on the second day, and increasing five times the first dose on the third day with acetyl salicylic acid dispersed in diffusion-controlled dosage forms with $\frac{D}{R^2} = 10^{-6}/s$. Apparent plasmatic volume = 10 l. 1. Dose of 1g. 2. Dose of 2 g.

For acetyl salicylic acid, this ratio is still larger, around 2.8 (Figure 8.12, curve 2).

- The plasma drug profiles, expressed in terms of plasma concentration, are drawn for acetyl salicylic acid in Figure 8.13 when the apparent plasmatic volume is 10 l, with the amount of drug in every dose of 1 g (curve 1) and of 2 g (curve 2).

REFERENCES

1. Heilmann, K., *Therapeutic Systems,* 2nd edition, Georg Thieme Verlag, Stuttgart, 1984, 20.
2. Caron, H.S. and Roth, H.P., Patients' cooperation with a medical regimen, *J. Am. Med. Assoc.,* 203, 120, 1968.
3. Ouriemchi, E.M. and Vergnaud, J.M., Assessment of the drug level in bronchial secretion with patient's non-compliance and oral dosage forms with controlled release, *Inflammopharmacol.,* 8, 267, 2000.

9 Drug Transfer in Various Tissues

NOMENCLATURE

A_l, A_{bs} Area of the lung in contact with the blood, with the mucus, respectively.

$Cl_{x,t}$ Drug concentration in the lung at position x and time t.

Cbs_t Uniform drug concentration in the mucus at time t.

Cp_t, C Uniform drug concentration in the blood at time t, drug concentration, respectively.

D Diffusivity of the drug in the tissue (lung, capillary wall).

GI (T)   Gastrointestine or gastrointestinal (tract).

h Convection coefficient in the mucus.

k_a; k_e Rate constant of absorption, of elimination (per h).

K Partition factor at the lung–blood interface.

K' Partition factor at the mucus–lung interface.

L; Ll Thickness of the lung.

L'; Lbs Thickness of the mucus.

M_t Amount of drug transferred in the blood up to time t.

Mbs_t Amount of drug transferred in the mucus up to time t.

t_r Time of full erosion of the erodible dosage form.

V_p Apparent plasmatic volume (l).

x, t Position, time.

Y Amount of drug available along the *GI*.

Z Amount of drug in the blood.

W Amount of drug eliminated.

Blood or plasma drug level has been used as an index of dose scheduling for therapeutics under the assumption that this drug level corresponds to the pharmacological effect of the drug. Conventional pharmacokinetic models have been widely applied to simulate the kinetics of drug levels in blood or plasma when the drug is delivered intravenously. Moreover, the kinetic information of the drug levels in blood and in various tissues or organs, such as the brain, cerebrospinal fluid, liver, muscles, and adipose tissue, has been evaluated in the case of intravenous drug delivery [1].

At first glance, the concept of tissue penetration may appear simplistic. However, tissue penetration regulates the clinical effectiveness and toxic potential of antibacterial agents and other drugs. In fact, knowledge of tissue distribution principles is essential before one can compare various agents [2]. Traditional pharmacokinetic data yield an antibiotic concentration-vs.-time curve that is descriptive of the drug behavior in plasma or serum [3, 4].

Actual measurements of concentration of aminoglycosides and β-lactams in tissues reveal uneven tissue distribution. Tissue–serum ratios of these compounds are thus less than 1:1 for most body sites, excluding the excretory organs [5, 6]. In contrast, some antibiotics have the ability to concentrate at body tissue sites, so that they were thought to be effective at lower dosages. These high concentrations could thus be associated with organ toxicity, as was the case with the high tissue–serum ratios of aminoglycosides, for which concentrations in the kidneys [7] or the perilymph [8] were related to the development of toxicity.

A consensus exists that antibiotics should be present in the extracellular fluid to be effective in treating bacterial infection. The concept stands that the effective concentration in tissue is directly related to the effective concentration in blood [2]. A variety of methods have been devised for determining the tissue penetration of antibiotics [9–12]. But these diverse methods often do not agree. First, it is essential to define the specific tissue in question, because antibiotics are not distributed evenly throughout the body. Furthermore, some tissues equilibrate rapidly with the vascular space; others, such as ascetic fluid, equilibrate more slowly. Some slowly equilibrating tissues may act as deposit sites for the antibiotic, such as kidney for aminoglycosides or bone for tetracyclines, maintaining persistently high concentration long after concentration in plasma has fallen toward zero [13]. However, concentrations in tissues vary depending on the method used to measure them [12, 14], because various factors affect the measurements: amount of blood in the tissue sample [11, 12]; sample desiccation; chemical degradation of the drug during processing [12]; destruction of cellular membranes with the release of intracellular proteins [15]; improper preparation of standards [10, 15]. Ideally, the tissue should be excised after the distribution phase of the drug has been completed and a steady state between the vascular space and the tissue has been achieved [16, 17]. Finally, the condition of the tissue at the time of excision should be known, because inflammatory reaction alters the physiochemical environment by varying the antibiotic uptake.

There are two types of mass transfer between the capillaries and the surrounding tissue. One type is the nearly instantaneous establishment of an equilibrium of the free drug concentration between the capillaries and the tissue, as for the liver [1]. The other type of mass transport through the capillary wall is the restricted passage between the capillary and the tissue, as in muscle; this restricted transport of drugs through the capillary wall is described in terms of transient diffusion [1]. A method for determining the diffusion coefficients of antibiotics through the extracellular tissue space was applied to rat brain tissue by considering a one-dimensional diffusion [18]; from the gradients of concentration measured with a voltmeter at different depths and times, the diffusivities were found to range from 10^{-7} to $2 \times 10^{-7} \, cm^2/s$.

Many interesting papers have been published on this subject when the drug was intravenously delivered, and it is hazardous to make a choice among them in this bird's-eye-view bibliography.

Penetration of ciprofloxacin in the cerebrospinal fluid (CSF) of patients with uninflamed meninges has shown that most antibiotics are insufficient when the blood–CSF barrier is intact, after intravenous delivery of the drug [19]. More recently, it has been shown that compliance is of utmost importance, and recommendations for starting and stopping treatment have been given for the therapy of epilepsies [20].

After its intravenous injection, the degree of penetration of an antibiotic into the bone and joint was found to be the important factor of its therapeutic efficiency, with isepamicin [21] or with levofloxacin into bone and synovial tissue [22], as well as with cefepime [23]. It was also found that ciprofloxacin should be used at high dosage, and that a single dose is not effective enough and not safe enough to prevent a posttraumatic osteitis [24].

9.1 DRUG TRANSFER INTO AND THROUGH THE LUNG AND BRONCHIAL MUCUS

9.1.1 SHORT BIBLIOGRAPHY

A drug can be transferred into the lungs and the bronchial secretion through two different means:

- By inhalation when the gaseous or atomized form of the drug is inhaled [25]
- Through either intravenous delivery or oral dosage forms

Only the second means of drug delivery is considered in this book, by oral dosage forms (either with immediate release or with sustained release) and by i.v. delivery.

Some papers are concerned with the modeling of the transfer of the drug into the lung and the mucus from the blood compartment, and these are briefly analyzed.

An interesting paper, previously cited and briefly described at the beginning of this chapter, is worth noting here [1]. By considering the drug balance and the resulting drug concentration–time profiles either in the blood or in various tissues, the system of ten simultaneous equations was solved, leading to the kinetics of the drug transferred in these parts of the body. The advantage of this method of calculation is that all the kinetics of drug concentration is obtained simultaneously. However, the diffusion of the drug through the capillary wall was oversimplified, and only a mean value of the concentration in the organs, such as muscle or adipose tissue, was obtained. Moreover, two drawbacks appear: the drug transport takes place in the lung tissue, which is so thick that it cannot be neglected, and there is no lag time between the concentrations–time profiles in the blood and in each organ. In fact, the presence of partial derivative equations expressing diffusion transport correctly would have made the mathematical treatment performed in this paper impossible.

The concentration–time of ciprofloxacin has been assessed in the blood compartment and in the lung tissue following oral administration of the drug. Immediate-release and erosion-controlled dosage forms have been examined. A numerical model, based on finite differences and taking into account all relevant data, has been built: the kinetics of drug release in the *GI* tract, drug absorption in the blood compartment and elimination, and the transient diffusion of the drug throughout the lung tissue. A partition coefficient for the drug at the tissue–blood interface has been considered to express the increase in the drug concentration at the tissue surface. The effect of the dose frequency and of the erosion rate of the dosage forms on the antibiotic concentration vs. time curves in the plasma and the lung tissue has been studied in detail [26].

The drug level has also been calculated in the plasma, the lung tissue, and bronchial secretion using ciprofloxacin as the drug, with erosion-controlled dosage forms. The numerical model built and tested for this calculation takes into account the following stages [27]:

- The kinetics of drug release along the GIT
- Absorption into the plasma and elimination from it
- The transient diffusion through the lung tissue
- Convection into the bronchial secretion

The effect of the patient's noncompliance was also emphasized [28].

This diffusion model was used for evaluating the transport of ciprofloxacin into the lung tissue [26], and the drug concentration–time history obtained either from calculation and experiment in the plasma and the lung were found to be in good agreement; the ciprofloxacin level in the lung tissue followed the ciprofloxacin plasma level, with a lag time resulting from the time necessary for the drug to diffuse through the lung [29].

The diffusion of the drug was also considered for evaluating the degree of penetration of the antibiotic into the lung tissue with moxifloxacin at a dose of 400 mg administered intravenously or orally once daily, and the results were correlated with microbiological data in order to estimate the clinical efficacy of the drug. This drug exhibits high penetration in lung tissue, with tissue concentrations far above the MIC 90 for most of the susceptible pathogens involved [30].

Other studies were made on the estimation of the drug concentrations either in the plasma or the lung and epithelial lining fluid for ceftazidime administered in continuous infusion to critically ill patients with severe nosocomial pneumonia [31]. In the same way, the concentrations of piperacillin–tazobactam in the plasma and lung tissue were determined in the case of steady-state drug concentrations [32].

In fact, as was shown in a review paper [33], after the evaluation of the concentration–time profile of the drug in the blood, it is possible to calculate the drug transfer by transient diffusion throughout the lung and by convection into the mucus, whatever the means of drug delivery in the body.

9.1.2 MATHEMATICAL AND NUMERICAL TREATMENT

As mentioned previously (Section 9.1.1), the four stages are considered in succession: the release of the drug from the dosage form along the GIT, the absorption into and elimination from the plasma, the diffusion through the lung tissue, and convection into the bronchial secretion.

9.1.2.1 Assumptions

The following assumptions are made in order to clarify the process:

- The kinetics of drug release along the GIT is very fast in the case of immediate-release dosage forms. In the case of sustained-release dosage forms, the process of the drug is controlled either by diffusion or by erosion, depending on the type of dosage form.

- The kinetics of drug release and the rate constant of absorption do not vary along the GIT time [34]. (In fact, this assumption is not necessary, provided that the variation is known).
- The stages of absorption and elimination of the drug are expressed by first-order kinetics with the rate constants of absorption and elimination.
- The rate constants of absorption and elimination, as well as the apparent volume of distribution, are determined from in vivo data, either in the case of the lung [35, 36] or of the mucus [37]. These values are kept constant in multidose treatment [38].
- The drug transport throughout the lung tissue is controlled by one-directional transient diffusion with a constant diffusivity. The typical representation of the lung is a tree, but because the distribution of the area and volume as a function of the radius of each component of this tree is not known, the lung is assumed to have a uniform thickness L. A partition factor K is introduced at the plasma–lung interface to account for the fact that the drug concentration on the lung tissue surface is K times that in the plasma [26], following the experimental data [35, 26].
- The transport of the drug from the lung tissue into the bronchial secretion (mucus) is controlled by convection with a convective transfer coefficient into the secretion next to the lung tissue surface. A partition factor K' is introduced [27], meaning that the drug concentration in the secretion is K' times that at the surface of the lung tissue in contact with the secretion. Because of the convection process, whose rate is much larger than the rate obtained with diffusion, the drug concentration is uniform in the bronchial secretion [39].
- The volumes of the lung tissue and of the bronchial secretion are much smaller than the volume of distribution, so they do not affect the plasma concentration. Thus, the problems of drug transfer into the plasma compartment and drug transfer into the lung and mucus are considered separately.

9.1.2.2 Evaluation of the Amount of Drug along the GI Tract and the Plasma

Depending on the type of dosage form, the amount of drug along the GIT is as follows:

- With immediate release, the total amount of drug M_{in} is available.
- With the sustained-release dosage form, the rate of drug released by the dosage form is $\frac{dM}{dt}$, whether it is controlled by diffusion or erosion.

The amount of drug in the GI, Y, is obtained by the following equation:

$$\frac{dY}{dt} = \frac{dM}{dt} - k_a Y \tag{9.1}$$

The amount of drug in the plasma Z is given by

$$\frac{dZ}{dt} = k_a Y - k_e Z \tag{9.2}$$

and the amount of drug eliminated W is

$$\frac{dW}{dt} = k_e Z \tag{9.3}$$

9.1.2.3 Transfer of the Drug into the Lung Tissue

The general equation expressing the mass balance for the drug at any time t is expressed by

$$M_t = Y_t + Z_t + W_t + A_l \int_0^L Cl_{x,t}\,dx + A_{bs}L_{bs}Cbs_t \tag{9.4}$$

where M_t is the amount of drug released from the dosage form up to time t; A_l and A_{bs} are the areas of the lung in contact with the plasma and the bronchial secretion, respectively; L and L_{bs} are the mean thicknesses of the lung and of the bronchial secretion, respectively; $Cl_{x,t}$ and Cbs_t represent the concentration of the drug in the lung at time t and position x, and the uniform concentration in the bronchial secretion at time t, respectively. See Figure 9.1.

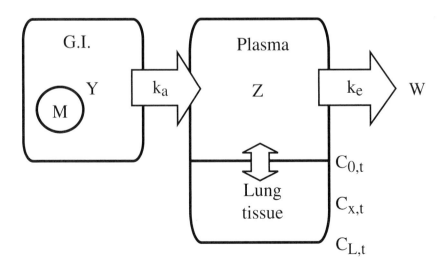

FIGURE 9.1 Scheme of the path followed by the drug from the *GI* tract to the lung. *Y*: amount of drug along the *GI* tract at time *t*. *Z*: amount of drug in the plasma at time *t*.

The equation of the unidirectional diffusion with constant diffusivity through the thickness of the lung tissue is

$$\frac{\partial C}{\partial t} = D \frac{\partial^2 C}{\partial x^2} \tag{9.5}$$

with the boundary conditions in contact with the plasma:

$$t > 0 \qquad x = 0 \qquad Cl_{0,t} = KCp_t \tag{9.6}$$

where Cp_t represents the uniform concentration of free drug in the plasma at time t, and K is the *partition factor* at the lung–plasma interface, meaning that the concentration of drug on the lung surface is K times that of the plasma at any time.

9.1.2.4 Transfer of Drug in the Bronchial Secretion

Depending on the viscosity of the material, the transfer of a substance into this material is controlled either by convection or by diffusion [39]. In the case of a liquid as the bronchial secretion, the transfer can be assumed to be controlled by convection:

$$-D \left(\frac{\partial C}{\partial x} \right)_{L,t} = h(Cl_{L,t} - K'Cbs_t) \tag{9.7}$$

This relation expresses that the rate at which the drug is brought to the interface by diffusion through the lung is constantly equal to the rate at which the drug enters the bronchial secretion [39]. The coefficient of convective transfer into the bronchial secretion is h, and its value depends on its viscosity. K' is the partition factor at the lung–bronchial secretion interface, meaning that the uniform concentration of the drug in the bronchial secretion is K' times that on the lung surface in contact with the liquid.

When no evaporation or excretion occurs from the bronchial secretion, the other boundary condition should be considered:

$$\left(\frac{\partial Cbs}{\partial x} \right)_{L'} = 0 \tag{9.8}$$

and the uniform concentration in the bronchial secretion is obtained by

$$Cbs_t = \frac{Mbs_t}{L'} \tag{9.9}$$

where Mbs_t is the amount of drug transferred into the bronchial secretion per unit area of lung–bronchial secretion interface at time t, and L' is the thickness of the bronchial secretion.

9.1.3 ABSTRACT OF THE EXPERIMENTAL PART

9.1.3.1 Experiments on Lung Tissue [35]

The pharmacokinetics of ciprofloxacin in the plasma and lung tissue at steady state were determined on 38 patients subjected to lung surgery, after they were given 500 mg of drug through an immediate-release dosage form. Plasma samples, two from each patient, were obtained and analyzed.

9.1.3.2 Experiments on Bronchial Secretion [37]

Eight adults (four females and four males) with a mean age of 24 years were given 500 mg oral ciprofloxacin every 8 h for a total of 10 days. Pharmacokinetic studies were performed on day 3 of therapy. Blood and sputum were taken at various times for drug analysis.

9.1.4 DATA COLLECTED FOR LUNG AND BRONCHIAL SECRETION

The pharmacokinetic parameters are collected in Table 9.1, as obtained from the authors [35, 37].

9.1.5 RESULTS OBTAINED WITH THE LUNG TISSUE

The scheme of the process is shown in Figure 9.1, with the blood compartment and the lung tissue through which the drug diffuses.

Good correlation between the experimental and calculated values is observed in Figure 9.2 for the profiles of drug concentration either in the plasma (curve 1) or in the lung tissue (curve 2). The mean theoretical value for the lung is only drawn in Figure 9.2, because the variation in the concentration is rather small, resulting from the very thin thickness of the tissue. Two values are used for the rate constant of elimination of the drug in calculating these profiles. Thus, again, the strong effect of this parameter is demonstrated.

TABLE 9.1
Pharmacokinetic Parameters of Ciprofloxacin

$k_a = 1.29/h$	$k_e = 0.12 - 0.2 - 0.32/h$	$V_p = 120$ litres
$D = 5 \times 10^{10}\,cm^2/s$	$L = 10\ \mu m$ (lung)	$K = 3.1$ (lung–serum interface)
$L' = 10\ \mu m$ (mucus)	$K' = 0.125$ (mucus–lung interface)	
$h = 2.7 \times 10^{-6}\,cm/s$	$\frac{hL}{D} = 5$	

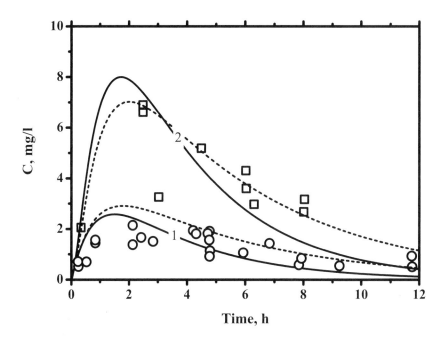

FIGURE 9.2 Profile of drug concentration in the plasma (1) and lung (2) with an immediate-release dosage form containing 500 mg ciprofloxacin. $k_a = 1.2/h$; $k_e = 0.32/h$ (solid line) and $k_e = 0.2/h$ (dotted line); $V_p = 120$ 1; $K = 3.1$; $D = 5 \times 10^{-10}\,cm^2/s$; $L = 10\,\mu m$.

Other drug profiles are drawn with an erosion-controlled dosage form containing 500 mg of ciprofloxacin, whose time of erosion is 24 h either for a single dose (Figure 9.3) or for repeated doses using the same dosage forms taken once a day (Figure 9.4).

Some comments are made regarding these figures:

- A short time lag appears in Figure 9.2 for the peak of the drug concentration reached in the lung tissue with respect to that in the blood. This lag is due to the time required for the drug to diffuse through the lung tissue, and especially to reach its midplane.
- A higher concentration is observed in the lung tissue than in the blood compartment, as determined by the experiments [35]. This fact necessitates that the model take a partition factor of 3.3 for the drug at the blood–lung tissue interface. It means that the drug concentration at the lung surface is 3.3 times more than that in the blood.
- Curve 2, the drug level in the lung tissue, represents the mean value through the thickness of the tissue. A slight change in the concentration of drug is obtained through the tissue thickness, resulting from its very thin thickness.
- Of course, it is possible to extend the numerical model in calculating the drug profile in both the blood and the lung tissue when sustained-release

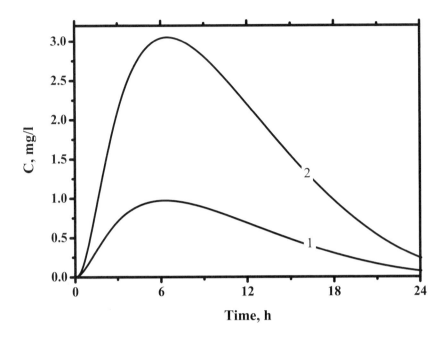

FIGURE 9.3 Profile of drug concentration in the plasma (1) and lung (2) with an erosion-controlled dosage form with a time of full erosion of 24 h, and 500 mg ciprofloxacin. $k_a = 1.2/h$; $k_e = 0.32/h$; $V_p = 120$ 1; $K = 3.1$; $L = 10$ μm; $t_r = 24$ h.

dosage forms are taken either in single dose (Figure 9.3) or via multidoses (Figure 9.4).

- Figure 9.4 shows that the steady state is promptly attained when sustained-release dosage forms whose release is controlled by erosion are taken once a day, with a more constant drug level than that obtained with immediate-release dosage forms.
- The experimental values are widely scattered. This dispersion results from the fact that various patients have been considered; only two tissue and blood samples were taken for each patient. Once again, the problem of intervariability plays an acute role, explaining the dispersion obtained through these tedious experiments, and thus in calculation two values were used for the rate constant of elimination.

9.1.6 RESULTS OBTAINED WITH THE BRONCHIAL SECRETION

The scheme of the process is shown in Figure 9.5, which is an extension of the process elaborated for drug transfer through the lung tissue with the addition of the convection of the drug into the highly viscous liquid of the mucus.

Figure 9.6 shows the profiles of drug concentration in the mucus and in the blood, as obtained by experiments [37] and calculation.

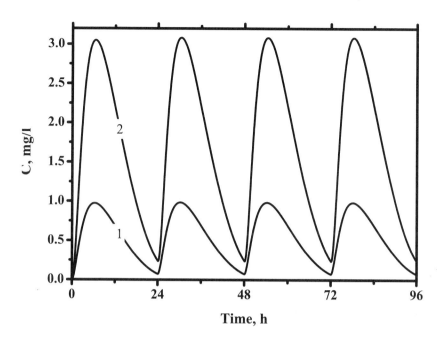

FIGURE 9.4 Profile of drug concentration in the plasma (1) and lung (2) with repeated doses taken once a day, each being an erosion-controlled dosage form with a time of full erosion of 24 h, and 500 mg ciprofloxacin. $k_a = 1.2/h$; $k_e = 0.32/h$; $V_p = 120$ l; $K = 3.1$; $L = 10$ µm; $t_r = 24$ h.

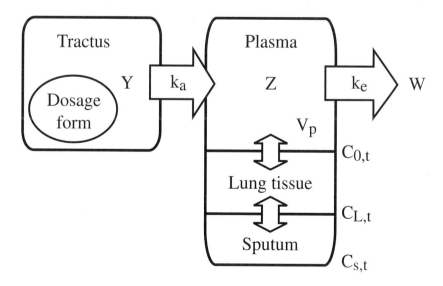

FIGURE 9.5 Scheme of the process of the drug transfer from the GIT to the mucus, through the blood and the lung tissue.

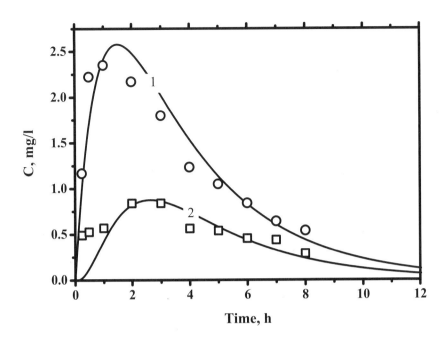

FIGURE 9.6 Profile of drug concentration in the plasma (1) and mucus (2) with an immediate-release dosage form containing 500 mg ciprofloxacin. $k_a = 1.2/h$ and $k_e = 0.2/h$; $V_p = 120$ 1; $K = 3.1$; $D = 5 \times 10^{-10}$ cm²/s; $L = 10$ μm; $L' = 10$ μm; $K' = 0.125$; h $= 2.7 \times 10^{-6}$ cm/s; and $\frac{hL}{D} = 5$. Reprinted from Inflammpharmacolo 6, pp. 321–337, 1998. With permission from Kluwer.

From these curves, the following conclusions can be drawn:

- Good agreement is obtained between the experimental and the calculated curves, proving the validity of the model. The maximum for each curve is also well evaluated either for the time or for the drug concentration in these two sites.
- The maximum of the drug level is attained after a longer time in the bronchial secretion (2.5 h) than in the blood (1.5 h). This lag time results from the time necessary for the drug to diffuse through the lung tissue before entering the mucus [27, 28].
- The experiments made with samples taken directly in the sputum and the blood were not so tedious as those made in the lung tissue, which explains the better profiles obtained.
- Of course, for calculation, the drug transport has been considered through the lung tissue and finally into the mucus, whereas only the drug profiles are drawn in the blood and the mucus in Figure 9.6. This is why Figure 9.7 shows the three profiles of the drug concentration in the blood (curve 1), in the lung tissue (curve 2), and in the mucus (curve 3). These profiles were obtained with an erosion-controlled dosage form having a time of full erosion of 24 h.
- Figure 9.8 shows the profiles obtained in the blood (1), the lung tissue (2), and in the mucus (3) as they are calculated in multidoses taken once

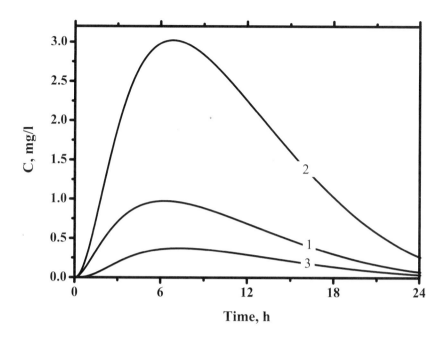

FIGURE 9.7 Profile of drug concentration in the plasma (1), in the lung (2), and in the mucus (3) with an erosion-controlled dosage form containing 500 mg ciprofloxacin. $k_a = 1.2/h$ and $k_e = 0.2/h$; $V_p = 120$ 1; $K = 3.1$; $D = 5 \times 10^{-10} \text{cm}^2/\text{s}$; $L = 10$ μm; $L' = 10$ μm; $K' = 0.125$; $t_r = 24$ h; $h = 2.7 \times 10^{-6}$ cm/s; and $\frac{hL}{D} = 5$.

　　　a day with erosion-controlled dosage forms having a time of full erosion
　　　of 24 h.
- 　It clearly appears that the drug concentrations in the blood and at the other
　　　two sites are lower with the sustained-release dosage form (Figure 9.7)
　　　than with its immediate-release counterpart (Figure 9.6), even if the
　　　amount of drug in each dosage form is the same. This is one of the main
　　　properties of the process of sustained release.

9.2　DRUG TRANSFER IN THE BLISTER FLUID

9.2.1　SHORT BIBLIOGRAPHICAL SURVEY

It is of great interest to get good knowledge of the diffusion of antibiotics in the extracellular fluid [18], because this is the obligatory path to infection loci. The diffusion of antibiotics has been studied in normal and pathological situations through various extracellular fluid samples, with the cerebrospinal fluid [19], prostatic fluid, bone and bradytrophic tissues (such as meniscus and other cartilage) [24], and bronchial secretions [29–32, 35–40]. More specifically, various experimental studies have been done on the penetration of the drug into the chemically induced blister fluid, which has been shown to be similar in composition to the exudates of a mild inflammatory reaction [41, 42].

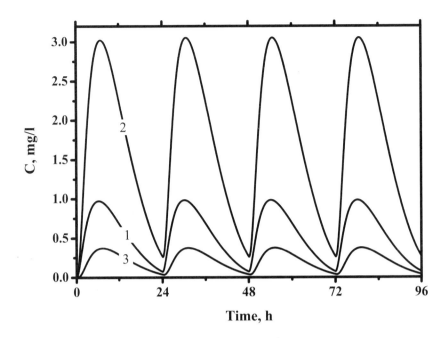

FIGURE 9.8 Profile of drug concentration in the plasma (1), in the lung (2), and in the mucus (3) with erosion-controlled dosage forms containing 500 mg ciprofloxacin taken once a day. $k_a = 1.2/\text{h}$ and $k_e = 0.2/\text{h}$; $V_p = 120$ 1; $K = 3.1$; $D = 5 \times 10^{-10}\,\text{cm}^2/\text{s}$; $L = 10\ \mu\text{m}$; $L' = 10\ \mu\text{m}$; $K' = 0.125$; $t_r = 24\text{h}$; $\text{h} = 2.7 \times 10^{-6}\,\text{cm/s}$; and $\frac{hL}{D} = 5$.

Little information exists in the literature about the diffusivity of antibiotics through the tissues. However, it has been shown that the penetration of antibiotics into tissues is governed by Fickian diffusion [43, 44]. The concentration of ciprofloxacin was measured either in the blood compartment or in the blister fluid of healthy male volunteers after they were given a single, immediate-release oral dose of 500 mg [45]. On the other hand, ciprofloxacin concentrations were determined in the plasma and cantharides-induced fluid of healthy volunteers after i.v. and oral administration [46]. Following these studies, the pharmacokinetics and the suction-induced blister fluid penetration of the drug were compared after a single oral dose and after multiple doses [38]. The effect of the infection on drug penetration has also been examined, as has the effect of drug penetration on the infection response [47].

An attempt was made for a theoretical approach by considering the fact that the transport of ciprofloxacin was driven by transient diffusion [18]. The equation of diffusion through a semi-infinite medium with a constant concentration on the surface was selected in order to simplify the problem. However, it must be said that this equation does not correspond with the finite thickness of the tissue and a constant concentration on the surface [39].

Finally, the process of drug transport in the blister fluid was studied by considering the transient diffusion through the capillary wall, with the drug concentration into the blood varying according to the pharmacokinetic parameters of the patient [48].

In order to simplify the problem, the transport was assumed to be unidirectional through the thickness of the wall, and the diffusivity constant. Various means of drug administration were examined, either oral dosage forms with immediate release or infusion at a constant rate. Because no analytical solution exists, a numerical model based on finite differences was built and tested by comparing the theoretical with the experimental values found in the literature [45, 46].

Because it was not possible to evaluate the thickness of the capillary wall, the diffusional transport through this tissue was defined by the diffusivity as a fraction of the square of this thickness, this ratio having the dimension of $(time)^{-1}$, in the same way as the rate constant of a first-order reaction, but leading to quite different implications.

9.2.2 MATHEMATICAL AND NUMERICAL TREATMENT

Two kinds of calculation are made: one concerned with the determination of the drug profile in the blood, and the other with the diffusional transport of the drug through the capillary wall to the blister fluid.

9.2.2.1 Assumptions

A few assumptions are made in order to make the process clear (Figure 9.9):

- The diffusional transport of the drug is unidirectional through the capillary wall, over a mean length L.
- The diffusion is Fickian with a constant diffusivity.

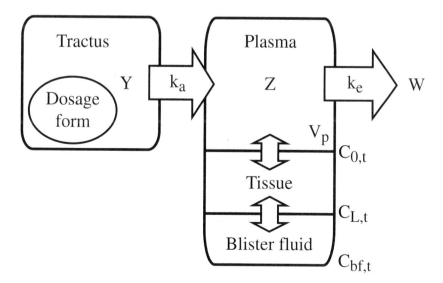

FIGURE 9.9 Scheme of the process of drug transfer in the blister fluid by diffusion through the capillary wall.

- The concentration of the drug on the surface of the wall in contact with the blood volume is the same as that in the plasma volume, at any time.
- The concentration of the drug on the surface of the wall in contact with the blister fluid is constantly equal to that of the blister fluid.

9.2.2.2 Mathematical Treatment of the Drug Transport

The profile of drug concentration in the blood is calculated in the same way as it was for the lung tissue, when the drug is administered through oral immediate-release dosage forms, and the equations are written again without additional comment.

The amount of drug in the GI, Y, is obtained by this equation:

$$\frac{dY}{dt} = \frac{dM}{dt} - k_a Y \qquad (9.1)$$

The amount of drug in the plasma Z is given by

$$\frac{dZ}{dt} = k_a Y - k_e Z \qquad (9.2)$$

and the amount of drug eliminated W is

$$\frac{dW}{dt} = k_e Z \qquad (9.3)$$

9.2.2.3 Administration of the Drug through Short Infusions

This problem was analyzed in depth in Chapter 2 (Section 2.4.1), with repeated doses at constant flow rate over a finite time. Only the principle is described here.

During the stage of infusion, the amount of drug in the plasma Z is given by

$$\frac{dZ}{dt} = K_0 - k_e Z \qquad (9.10)$$

whose solution allows one to obtain the amount of drug in the blood at the end of the time of infusion t_i; K_0 is the constant rate of drug delivered by the infusion during this time.

$$Z_{ti} = \frac{K_0}{k_e}[1 - \exp(-k_e t_i)] \qquad (9.11)$$

After the infusion has been stopped, the amount of drug in the blood decreases according to the simple equation, up to the time of the second infusion:

$$Z_t = Z_{ti} \exp[-k_e(t - t_i)] \qquad (9.12)$$

9.2.2.4 Diffusion of the Drug through the Capillary Wall

The equation of diffusion through the wall of thickness L is

$$\frac{\partial C}{\partial t} = D \frac{\partial^2 C}{\partial x^2} \tag{9.5}$$

and the initial conditions express that there is no drug in the wall:

$$t = 0 \qquad 0 < x < L \qquad C = 0 \tag{9.13}$$

The boundary conditions are for the surface in contact with the plasma:

$$t > 0 \qquad x = 0 \qquad C_{0,t} = C_{plasma,t} \tag{9.14}$$

and for the surface in contact with the blister fluid:

$$t > 0 \qquad x = L \qquad C_{L,t} = \text{blister fluid concentration} \tag{9.15}$$

9.2.2.5 Numerical Analysis

Because no analytical solution exists for the problem of fluctuating concentrations on each surface, a numerical method with finite differences is built and used to resolve the problem.

9.2.3 THEORETICAL AND EXPERIMENTAL RESULTS

Two kinds of results should be compared simultaneously in order to test the validity of the model:

- The drug concentration in the blood
- The drug concentration obtained in the blister fluid

The experimental and calculated drug profiles obtained in the blood and in the blister fluid, with either oral immediate-release dosage form or infusion, are thus examined [46, 48].

The drug profiles in the blood (1) and the blister fluid (2) are drawn in Figure 9.10 for a single oral dose with immediate release, in Figure 9.11 with the same dosage form delivered in multidoses, and in Figure 9.12 with repeated infusions, using the experimental values shown in Table 9.2 [46]. These profiles lead to the following comments:

- Good agreement is obtained between the calculated and experimental values, either in the blood or in the blister fluid, for the oral dosage form with immediate release, as shown in Figure 9.10. This is due to the fact that the method used in sampling the blister fluid for analysis is highly

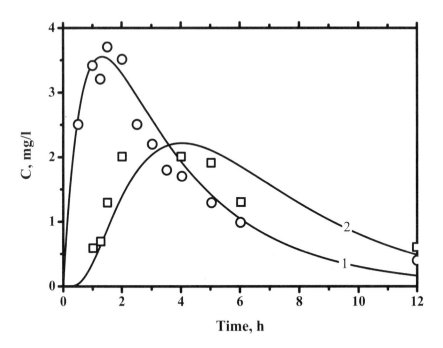

FIGURE 9.10 Profiles of drug concentration in the blood (1) and in the blister fluid (2) obtained with an oral dosage form with immediate release containing 750 mg ciprofloxacin. $k_a = 1.5/h$ and $k_e = 0.3/h$; $V_p = 140$ 1; $\frac{D}{L^2} = 0.2/h$. Reprinted from Inflammpharmacolo 6, pp. 321–337, 1998. With permission from Kluwer.

convenient and efficient. There was also little intervariability among the six healthy volunteers who were administered the drug.

- With the oral administration in Figure 9.10, a maximum of the drug concentration is attained at 1.33 h in the blood and at around 4 h in the blister fluid. Thus, the drug concentration in the blister fluid lags far behind the corresponding one in the blood, due to the process of diffusion through the capillary wall.
- As shown in Figure 9.11, with repeated doses at the rate of three times a day using the same oral dosage form with immediate release as in Figure 9.10, the drug concentration in the blister fluid alternates between peaks and troughs in the same way as that in the blood, with a time lag of around 2.6 h.
- Of course, resulting from the process of diffusion with no partition factor (which could increase the drug concentration in the same way as in the lung), the profile of drug concentration in the blister fluid is flatter than that in the blood.
- Typical drug profiles are shown in Figure 9.12, either for the blood or for the blister fluid, when the drug is delivered through a 1 h infusion given three times a day. These profiles are similar to those obtained with oral dosage forms, with a difference in the concentration level (which is higher

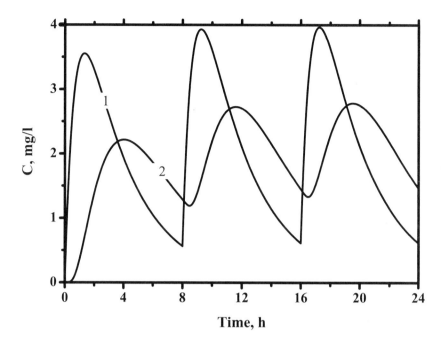

FIGURE 9.11 Profiles of drug concentration in the blood (1) and in the blister fluid (2) obtained with repeated oral dosage form with immediate release containing 750 mg ciprofloxacin, taken three times a day. $k_a = 1.5/h$ and $k_e = 0.3/h$; $V_p = 140$ 1; $\frac{D}{L^2} = 0.2/h$.

with i.v. delivery). This results from the effect of the first-pass hepatic, which reduces the drug level in the case of oral delivery [46].

- The values of the parameters of diffusion necessary for fitting the theoretical and experimental curves are the same, whatever the means of drug delivery: around 0.2/h for the ratio $\frac{D}{L^2}$. Note that this ratio has the same dimension as a rate constant of absorption, but with a very different behavior: in the case of diffusion the drug is transported from the higher concentration to the lower, which allows the drug concentration to alternate between peaks and troughs in the blister fluid in following the plasma drug profile.

9.3 DRUG TRANSFER INTO THE ENDOCARDITIS

9.3.1 SHORT BIBLIOGRAPHICAL SURVEY

Bacterial endocarditis is a difficult infection to cure, due to poor penetration of antibiotics into infected vegetations, altered metabolic state of bacteria within the lesion, and absence of adequate host-defense cellular response, which could cooperate with antibiotic action. The contribution of animal models to a better understanding

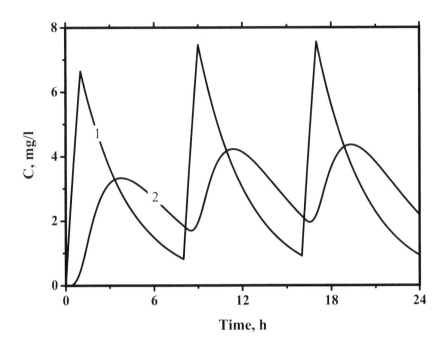

FIGURE 9.12 Profiles of drug concentration in the blood (1) and in the blister fluid (2) obtained with repeated infusions containing 400 mg ciprofloxacin over 1 h, delivered three times a day. $k_a = 1.5/h$ and $k_e = 0.3/h$; $V_p = 140$ 1; $\frac{D}{L^2} = 0.2/h$.

of the pathophysiology of the infection and to the definition and improvement of therapeutic regimens of endocarditis in humans is still of great importance, due to the difficulties encountered in clinical trials [49]. In fact, although it is usually effective to treat a patient using antibiotics when the bacterial infection is in the blood compartment, such treatment becomes difficult in the case of a vegetation infected by bacteria [50]. The striking example of the persistence of *Streptococcus mitis* in such a situation has been observed over some 25 days after antibiotic therapy [51]. This well-known fact appears in patients with endocardial vegetations containing bacteria, and the aggressive treatment of infectious endocarditis requires high doses of antibiotics to achieve an effective cure [52]. Such treatment is needed because the antibiotics present in the blood compartment should diffuse within the vegetation before exerting their activity against those bacteria located in the vegetation. This diffusional stage

TABLE 9.2
Pharmacokinetic and Diffusion Parameters of Ciprofloxacin

$k_a = 1.5/h$	$k_e = 0.3/h$	$V_p = 140$ 1	$\frac{D}{L^2} = 0.2/h$

of the drug is controlled by transient radial diffusion and is time consuming, because the time necessary for the drug to reach the center of this sphere is proportional to the square of the radius of the vegetation [39]. Thus, the treatment of endocarditis often requires prolonged antibiotic therapy, and it is important to know the terms of adequate therapy; individualized drug dosage regimens have made therapy possible even in patients with impaired renal function [52].

The knowledge brought with the typical characteristics of diffusion through the spherical endocarditis enabled the medical doctors to define clearly the special features of the therapy, with the amount of drug and the duration of the drug delivery [53], and these were applied to a successful treatment of various patients in the hospital by using repeated infusions with the calculated dose and duration of therapy [54].

On the other hand, an in vitro method was developed in order to evaluate precisely the main characteristic of the diffusion, such as the diffusivity of a drug [55].

9.3.2 MATHEMATICAL AND NUMERICAL TREATMENT OF THE DRUG TRANSPORT

9.3.2.1 Assumptions

The following assumptions are made to clarify the problem:

- The drug profile in the blood is obtained with discontinuous infusions, repeated twice a day, each dose of 500 mg being delivered with a constant rate over 0.5 h.
- The vegetation, spherical in shape, is in contact with the blood.
- The drug transport through (into and out of) the sphere is controlled by radial diffusion, with a finite coefficient of convection at the blood–vegetation interface.

9.3.2.2 MATHEMATICAL TREATMENT

The drug profile in the blood is calculated using Equation 9.10 through Equation 9.12.
The equation of the radial diffusion is [39]:

$$\frac{\partial C}{\partial t} = \frac{1}{r^2}\frac{\partial}{\partial r}\left[Dr^2\frac{\partial C}{\partial r}\right] \tag{9.16}$$

The initial condition expresses the fact that the vegetation is free of drug:

$$t = 0 \qquad 0 < r < R \qquad C = 0 \tag{9.17}$$

The boundary condition expresses the fact that at the blood–vegetation interface, the rate of drug transfer by convection is constantly equal to the rate transfer by diffusion into the sphere.

$$-D\frac{\partial C}{\partial r} = h(C_{blood,t} - C_{R,t}) \tag{9.18}$$

Because of the variable concentration of the drug in the blood, mathematical treatment is not feasible, and a numerical method should be used, with a constant increment of time and consideration of the spherical penetration along the increment of the radius [39].

9.3.3 RESULTS WITH THE DRUG PROFILES IN THE VEGETATION

The profiles of drug concentration are drawn in Figure 9.13 (with the ratio $\frac{D}{R^2} = 10^{-6}$/s) and in Figure 9.14 (with the ratio $\frac{D}{R^2} = 5.10^{-6}$/s) in the blood (1), on the surface of the vegetation (2), at half its radius (3), and at the center of the sphere (4).

These curves lead to some comments of interest for therapy:

- The drug diffuses slowly along the radius of the spherical vegetation, and the profiles at different positions in the sphere are quite different.
- The effect of the characteristics of the vegetation with its radius and the value of the diffusivity of the drug through it is of great importance on the penetration of the drug. Thus with the lower value of the ratio $\frac{D}{R^2}$, in Figure 9.13, the steady state in not attained after 72 hours of infusion, whereas it is attained after the second dose with the higher value of this ratio (Figure 9.14).

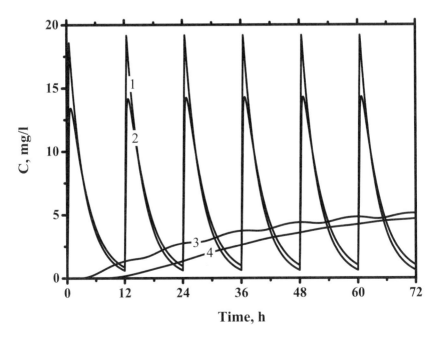

FIGURE 9.13 Profiles of drug concentration in the blood (1) and in various positions in the vegetation: 2. on the surface; 3. at the midradius; 4. in the center; with the value of the ratio $\frac{D}{R^2} = 10^{-6}$/s, with amikacin delivered through infusions of 500 mg over 30 min twice a day. $k_e = 0.30$/h; $V_p = 25$ 1.

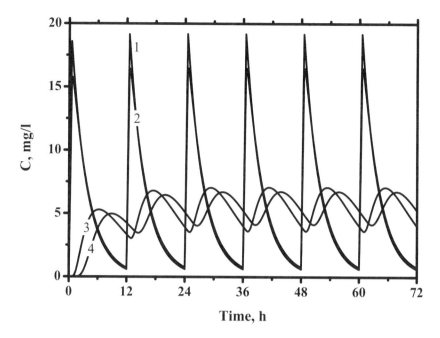

FIGURE 9.14 Profiles of drug concentration in the blood (1) and in various positions in the vegetation: 2. on the surface; 3. at the midradius; 4. in the center; with the value of the ratio $\frac{D}{R^2} = 5.10^{-6}/s$, with amikacin delivered through infusions of 500 mg over 30 min twice a day. $k_e = 0.30/h$; $V_p = 25$ 1.

- The drug concentration in various positions of the vegetation alternates between peaks and troughs in the same way as in the blood compartment. This fact is especially true in Figure 9.14, with the larger value of the ratio $\frac{D}{R^2}$.
- The ratio $\frac{D}{R^2}$ is of great concern, because either the diffusivity or the radius of the vegetation is difficult to determine with precision. Thus, instead of having two unknowns with the diffusivity and the radius, there is only one.
- The obvious statement holds: the smaller the radius of the vegetation and the larger the diffusivity of the drug through it, the faster the drug transport into the vegetation.
- It is worth noting again that, because the process is controlled by diffusion, the time of penetration in the vegetation is proportional to the square of its radius.

REFERENCES

1. Chen, C.N. et al., Pharmacokinetic model for salicylate in cerebrospinal fluid, blood, organs and tissues, *J. Pharm. Sci.*, 67 (1), 38, 1978.
2. Nix, D.E. et al., Antibiotic tissue penetration and its relevance: models of tissue penetration and their meaning, *Antimicrob. Agents Chemother.*, 35 (10), 151, 1991.

3. Drusano, G.L. et al., Integration of selected pharmacologic and microbiologic properties of three new β-lactam antibiotics: a hypothesis for rational comparison, *Rev. Infect. Dis.*, 6, 357, 1988.

4. Schentag, J.J., Swanson, D.J., and Smith, I.L., Dual individualization: antibiotic dosage calculation from the integration of in vitro pharmacodynamics and in vivo pharmacokinetics, *J. Antimicrob. Chemother.*, 15 (Suppl. A), 47, 1985.

5. Rolinson, G.N., Tissue penetration of antibiotics, *J. Antimicrob. Chemother.*, 13, 593, 1984.

6. Ryan, D.M. and Cars, O., Antibiotic assays in muscle: are conventional tissue levels misleading as indicators of antibacterial activity? *Scand. J. Infect. Dis.*, 12, 307, 1980.

7. Schentag, J.J. and Gengo, F.M., Principles of antibiotic tissue penetration and guidelines for pharmacokinetic analysis, *Med. Clin. North Am.*, 66, 39, 1982.

8. Stupp, H. et al., Kanamycin dosage and levels in ear and other organs, *Arch. Otolaryngol.*, 86, 63, 1967.

9. Bergan, T., Pharmacokinetics of tissue penetration of antibiotics, *Rev. Infect. Dis.*, 3, 45, 1981.

10. Gerding, D.N. and Peterson, L.R., Extravascular antibiotic penetration: skin and muscle, in *Antimicrobial Therapy*, Ristuccia, A.M. and Cunha, B.A., Eds., Raven Press, New York, 1984, 389.

11. Whelton, A. and Stout, R.L., An overview of antibiotic tissue penetration, in *Antimicrobial Therapy*, Ristuccia, A.M. and Cunha, B.A., Eds., Raven Press, New York, 1984, 365.

12. Wise, R., Methods for evaluating the penetration of β-lactam antibiotics into tissues, *Rev. Infect. Dis.*, 8 (Suppl. 3), 5303, 1986.

13. Schentag, J.J. et al., Gentamicin disposition and tissue accumulation on multiple dosing, *J. Pharmacokinet. Biopharm.*, 5, 559, 1977.

14. Ryan, D.M. et al., Simultaneous comparison of three methods for assessing ceftazidime penetration into extravascular fluid, *Antimicrob. Agents Chemother.*, 22, 995, 1982.

15. Cars, O. and Ogren, S., Antibiotic tissue concentrations: methodological aspects and interpretation of results, *Scand. J. Infect. Dis.*, 44, 7, 1985.

16. Gengo, F.M. and Schentag, J.J., Rate of methicillin penetration into normal heart valves and experimental endocarditis lesions, *Antimicrob. Agents Chemother.*, 21, 456, 1982.

17. Van Etta, L.L. et al., Effect of the ratio of surface area to volume on the penetration of antibiotics into extravascular spaces in an in vitro model, *J. Infect. Dis.*, 146, 423, 1982.

18. Meulemans, A. et al., Measurement and clinical and pharmacokinetic implications of diffusion coefficients of antibiotics in tissues, *Antimicrob. Agents Chemother.*, 33, 1286, 1989.

19. Nau, R. et al., Penetration of ciprofloxacin into the cerebrospinal fluid of patients with uninflamed meninges, *J. Antimicrob. Chemother.*, 25, 965, 1990.

20. Maire, P., The brain in rapid technological progress: epileptology for the general practitioner, *Schweiz Rundsch Med. Prax*, 92, 72, 2003.

21. Boselli, E. et al., Diffusion of isepamicin into cancellous and cortical bone tissue, *J. Chemother.*, 14, 361, 2002.

22. Rimmele, T. et al., Diffusion of levofloxacin into bone and synovial tissues, *J. Antimicrob. Chemother.*, 53, 533, 2004.

23. Breilh, D. et al., Diffusion of cefepime into cancellous and cortical bone tissue, *J. Chemother.*, 15, 134, 2003.

24. Wacha, H. et al., Concentration of ciprofloxacin in bone tissue after single parenteral administration to patients older than 70 years, *Infection*, 18, 173, 1990.

25. Falson-Rieg, F., Faivre, V., and Pirot, F., Microspheres pour administration nasale; Dispositifs d'administration pulmonaire, in *Nouvelles formes médicamenteuses,* Tec & Doc Eds, Paris, 2004, Chapters 8 and 10.

26. Saidna, M., Ouriemchi, E.M., and Vergnaud, J.M., Assessment of antibiotic levels in lungs tissue with erosion-controlled dosage forms, *Eur. J. Drug Metab. Pharmacokinet.,* 22 (3), 237, 1997.

27. Saidna, M., Ouriemchi, E.M., and Vergnaud, J.M., Calculation of the antibiotic level in the lung tissue and bronchial secretion with an oral controlled-release dosage form, *Inflammopharmacol.,* 6, 321, 1998.

28. Ouriemchi, E.M. and Vergnaud, J.M., Assessment of the drug level in bronchial secretion with patient's non-compliance and oral dosage forms with controlled release, *Inflammopharmacol.,* 8 (3), 267, 2000.

29. Breilh, D. et al., Mixed pharmacokinetic population study and diffusion model to describe ciprofloxacin lung concentrations, *Comput. Biol. Med.,* 31 (3), 147, 2001.

30. Breilh, D. et al., Diffusion of oral and intravenous 400 mg once-daily moxifloxacin into lung tissue at pharmacokinetic steady-state, *J. Chemother.,* 15 (6), 558, 2003.

31. Boselli, E. et al., Plasma and lung concentrations of ceftazidime administered in continuous infusion to critically ill patients with severe nosocomial pneumonia, *Intensive Care Med.,* 30 (5), 989, 2004.

32. Boselli, E. et al., Steady-state plasma and intrapulmonary concentrations of piperacillin/tazobactam 4 g/0.5 g administered to critically ill patients with severe nosocomial pneumonia, *Intensive Care Med.,* 30 (5), 976, 2004.

33. Vergnaud, J.M., Use of polymers in pharmacy for oral dosage forms with controlled release, *Recent Res. Devel. Macromol. Res.,* 4, 173, 1999.

34. Amidon, G.L. et al., A theoretical basis for a biopharmaceutical drug classification: the correlation of in vitro drug product dissolution and in vivo bioavailability, *Pharm. Res.,* 12, 413, 1995.

35. Breilh, D., Diffusion de la ciprofloxacine et de son principal métabolite la désethyl-ciprofloxacine dans le parenchyme pulmonaire humain. Etude des modélisations plasmatiques et simulation des concentrations pulmonaires de ciprofloxacine selon un modèle physiologique–pharmacocinétique, *Diplôme d'Etudes Approfondies,* Pessac, France, 1994.

36. Smith, B.R. and Frocq, J.L., Bronchial tree penetration of antibiotics, *Chest,* 6, 904, 1983.

37. Smith, M.J. et al., Pharmacokinetics and sputum penetration of ciprofloxacin in patients with cystic fibrosis, *Antimicrob. Agents Chemother.,* 30, 614, 1986.

38. Le Bel, M., Vallée, F., and Bergeron, M., Tissue penetration of ciprofloxacin after single and multiple doses, *Antimicrob. Agents Chemother.,* 29, 501, 1986.

39. Vergnaud, J.M., The diffusion equations and basic considerations, in *Controlled Drug Release of Oral Dosage Forms,* Ellis Horwood Eds, Chichester, U.K., 1993, Chapter 1.

40. Saux, P. et al., Penetration of ciprofloxacin into bronchial secretions from mechanically ventilated patients with nosocomial bronchopneumonia, *Antimicrob. Agents Chemother.,* 38, 901, 1994.

41. Wise, R. et al., The influence of protein binding upon tissue fluid levels of six β-lactam antibiotics, *J. Infect. Dis.,* 142, 77, 1980.

42. Wise, R. and Donovan, I.A., Tissue penetration and metabolism of ciprofloxacin, *Am. J. Med.,* 82 (suppl 4 A), 103, 1987.

43. Bergan, T., Pharmacokinetics of tissue penetration of antibiotics, *Rev. Infect. Dis.,* 3, 45, 1981.

44. Bergeron, M., Tissue penetration of antibiotics, *Clin. Biochem.,* 19, 90, 1986.

45. Crump, B., Wise, R., and Dent, J., Pharmacokinetics and tissue penetration of cipro-floxacin, *Chemotherapy*, 24, 784, 1993.

46. Catchpole, C. et al., The comparative pharmacokinetics and tissue penetration of single dose ciprofloxacin 400 mg iv and 750 mg po, *J. Antimicrob. Chemother.*, 33, 103, 1994.

47. Nix, D.E. et al., Antibiotic tissue penetration and its relevance: impact of tissue penetration on infection response, *Antimicrob. Agents Chemother.*, 35, 1953, 1991.

48. Bakhouya, A., Saïdna, M., and Vergnaud, J.M., Calculation of the blister-fluid history with ciprofloxacin administered orally or by infusion, *Int. J. Pharm.*, 146, 225, 1997.

49. Carbon, C., Animal models of endocarditis, *Int. J. Biomed. Comput.*, 36, 59, 1994.

50. Rosca, I.D. and Vergnaud, J.M., Modelling the process of diffusion and antibacterial activity of antibiotics in endocardial vegetations, *Pharm. Sci.*, 1, 391, 1995.

51. Voiriot, P., Weber, M., and Gerard, A., Persistence of *Spectrococcus mitis* in aortic vegetation after 25 days of penicillin–netilmicin combination therapy, *N. Engl. J. Med.*, 318, 1067, 1988.

52. Maire, P. et al., Computation of drug concentrations in endocardial vegetations in patients during antibiotic therapy, *Int. J. Biomed. Comput.*, 36, 77, 1994.

53. Confesson, M.A. et al., Traitement antibiotique de l'endocardite infectieuse: apport de la simulation des concentrations d'antibiotiques dans les végétations cardiaques, *Soc. de Pharmacie de Lyon,* 177, 17 décembre 1992.

54. Confesson, M.A. et al., Concentrations calculées en aminoside dans des végétations d'endocardites, *Thérapie,* 49, 27, 1994.

55. Senoune, A. et al., Intravegetation antimicrobial distribution in endocarditis: a numer-ical model and establishment of the conditions for an in-vitro test, *Int. J. Biomed. Comput.,* 36, 69, 1994.

10 Transdermal Therapeutic Systems

NOMENCLATURE

C Concentration of the drug.

$C_{x,t}$ Concentration of the drug at position x and time t.

C_0 Constant concentration of drug on the surface of the skin.

D Diffusivity of the drug in the skin, in the TTS in Section 10.5.1 and Section 10.5.2.

h Coefficient of drug transfer after removal of the TTS.

k_a Rate constant of absorption of the drug (per h).

k_e Rate constant of elimination of the drug (per h).

L Thickness of the skin (cm).

M_t Amount of drug transferred through the skin after time t.

n Integer in series.

TTS   Transdermal therapeutic system, patch.

t Time.

$t_{0.5}$ Half-life of the drug (h).

V_p Plasmatic volume (l).

x Abscissa taken along the thickness of the skin, for calculation.

X Amount of drug at the internal surface of the skin.

Y Amount of drug in the blood compartment.

Z Amount of drug eliminated.

In contrast to infusion systems, transdermal therapeutic systems deliver the drug systemically—in theory, pertaining to the whole body, not specifically to any one part—without the need of a needle to inject the drug into the patient's circulatory system. The terms defined by the FDA are as follows:

- Patch (for the common name)
- Patch with extended release
- Patch with extended release electrically controlled

The term *transdermal therapeutic system* (TTS) is also widely used.

10.1 GENERAL MECHANISMS OF DRUG DELIVERY

The structure of the TTS shown in Figure 10.1 is made of various parts [1–3]. It consists of the following:

- A covering membrane in contact with the drug reservoir protecting it

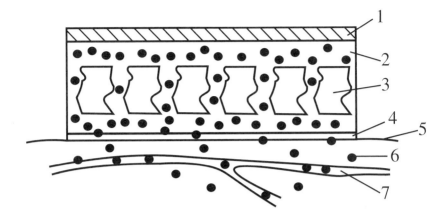

FIGURE 10.1 Scheme of a general transdermal therapeutic system applied to the skin. 1. Covering membrane. 2. Drug reservoir. 3. Micropore membrane controlling drug delivery 4. Adhesive contact surface. 5. Surface of skin. 6. Drug molecules. 7. Capillary.

- The drug reservoir, containing the drug in liquid state or in dissolved state in a liquid or a gel
- A delivery-control element
- An adhesive layer to keep the TTS in contact with the skin
- A protective foil, which should be removed from the system before application to the skin surface

The outward appearance of a TTS is a plaster or patch. The TTS, or patch with extended release, may be as thin as 150 μm and can cover an area of 5 to 20 cm^2 of the skin surface. This patch should be applied to a part of the body where the skin has a constant thickness and is supplied by a relatively constant and high blood flow. Some examples are behind the ear, on the upper arm, and on the chest.

The process of drug delivery by the TTS is mainly the diffusion of the drug through the skin from the TTS–skin interface to the subcapillary plexus. Diffusion starts as soon as the TTS is applied to the skin: first under transient conditions, because it takes some time for the drug to cross the skin and to saturate the binding sites of the skin. The transient condition is followed by a steady state, which is attained with a constant rate of drug delivery associated later with a constant drug concentration in the plasma. Finally, when the TTS is removed, the plasma drug concentration falls at a rate that depends on the drug's rate of elimination.

The following parameters intervene in the TTS process:

- The solubility of the drug in the skin, and especially on the stratum corneum surface, which depends on the nature of the drug and the epidermis.
- The diffusion of the drug through the skin, which depends on the nature of the drug and, to some extent, on the characteristics of the skin.

- The pharmacokinetic parameters of the drug, such as the rate constant of absorption into the blood and the rate constant of elimination from it.
- The amount of drug contained in the reservoir, which is crucial to the length of time over which the TTS remains functional.
- The TTS itself exhibits a limited rate of release of the drug, but generally this rate is much higher than the rate of drug transport through the skin.

10.2 OVERVIEW OF THE SKIN AND ITS ROLE

10.2.1 MAIN CHARACTERISTICS AND PROPERTIES OF THE SKIN

The skin is the largest organ of the body and one of its toughest tissues. Self-healing, water resistant, temperature regulating, infection repelling and, despite all that, highly sensitive, the skin basically consists of two layers:

- **Epidermis:** The outer layer is itself divided into four layers. The innermost layer, the germinative or Malpighian layer, is composed of cells constantly replicating and rising through the next two layers, the stratum granulosum and the stratum lucidum, whose keratin content increases by degrees to become the outer layer, called the stratum corneum. This outer, cornified layer consists entirely of dead cells with keratin instead of cytoplasm, and these dead cells continually flake off.
- **Dermis:** This layer (also called the corium) lies beneath the epidermis. The second of the two layers of the skin, the dermis is under the epidermis but on top of the thin basal membrane. In its thick single stratum of connective tissue are sensory nerve endings, blood capillaries, lymph vessels, tiny muscle fibers, sweat glands, sebaceous glands, and hair follicles. Beneath, there is a layer of fat that can be used as a store of food and water.

The thickness of the stratus corneum is between 10 and 20 μm; the thickness of the epidermis ranges from 0.04 to 1.5 mm, and the thickness of the dermis varies from 1 to 4 mm, depending upon the part of the body.

A part of the skin's function is to provide a protective barrier against invasion from microorganisms and toxic materials, as well as to prevent the loss of water or other physiological elements from the body. Thus, the skin is relatively impermeable to most substances; for that reason, it was long considered an inappropriate site for the administration of drugs for systemic use.

A prerequisite of transdermal drug absorption is the penetration of the drug through the layers of the epidermis and into the underlying tissue of the dermis, where the drug can gain access to the capillary vessels. Through in vitro tests, it has been shown that the stratum corneum is the principal barrier against many substances.

On the whole, it can be said that the stratum corneum is impermeable to hydrophilic molecules and slightly permeable to lipophilic molecules. The other parts of the skin, the living epidermis and the dermis, are rather permeable to hydrophilic molecules.

The permeability of the stratum corneum can be increased either by stripping or (much better) by using lipophilic enhancers of diffusion. Amphiphile molecules, both hydrophilic and lipophilic, could be another solution.

10.2.2 CLINICAL POSSIBILITIES OFFERED BY THE SKIN

From the characteristics briefly summarized and its position in the body, the skin becomes a way to attain systemic circulation, offering various advantages:

- The various phenomena affecting the gastrointestinal absorption of oral dosage forms are circumvented.
- Because the liver is bypassed, the first-pass hepatic does not play the role produced with oral dosage forms, and the drug metabolism is reduced to that which could intervene through the skin.
- Compounds with a narrow therapeutic index can be used more easily.
- Drugs with a short biological half-life can be employed.
- Treatment can be started and terminated at any time by applying the patch to or removing it from the skin.
- Patient compliance is easily improved.
- Theoretically, at least, there is no problem with time of delivery, as is imposed by the gastrointestinal tract time. Thus a steady state of drug delivery can be envisaged, with a constant plasma drug level.
- The risk of overdosage when the drug enters the circulation can be avoided, reducing the side effects associated with this problem.

10.3 CALCULATION OF THE DRUG TRANSPORT

10.3.1 ASSUMPTIONS

The following assumptions are made not only for calculation, but also for clear description of the process:

- The TTS can maintain a constant drug concentration on the surface of the skin.
- Contact at the TTS–skin interface is perfect, so that the drug concentration on the surface of the skin reaches the constant value as soon as the patch is put in contact with the skin. (This assumption is not absolutely necessary, but it is difficult to evaluate the coefficient of mass transfer h in this case.) Nevertheless, the effect of the pressure has been found to be of concern in this coefficient [4].
- The drug dissolves on the surface of the skin; according to the respective values of this solubility and the concentration of the drug in the TTS, a partition factor may appear.
- The drug diffuses through the various parts of the skin; in order to simplify, the skin is homogeneous and the diffusivity is constant. (It is possible to consider a multilayer skin with various diffusivities, but so far the experimental values are not known.)

- In the innermost part of the skin, the drug is absorbed into the capillaries. This stage of absorption follows a first-order kinetics with the rate of absorption, in the same way as it is absorbed into the blood compartment when it is taken orally, because these two processes are similar [3, 5–8].
- The drug is finally eliminated from the blood compartment with the first-order kinetics and the rate constant of elimination.

10.3.2 Calculation of the Drug Transfer into the Blood

As shown in Figure 10.2, the drug concentration is constant on the external surface of the skin, and this fact becomes one of the two boundary conditions. The other boundary condition is written by considering that the concentration of the drug at the innermost skin is very low, because the drug present there is absorbed into the capillaries. Before deposition of the patch, the skin is free of drug, leading to the initial condition.

Thus, the initial and boundary conditions for the skin of thickness L playing the role of a membrane are as follows:

$$t = 0 \quad 0 < x < L \quad C = 0 \tag{10.1}$$

$$t > 0 \qquad x = 0 \quad C = C_0 \tag{10.2}$$

$$x = L \quad C = \varepsilon$$

The one-directional transient diffusion through the skin with constant diffusivity is expressed by the partial derivative equation:

$$\frac{\partial C}{\partial t} = D \frac{\partial^2 C}{\partial x^2} \tag{10.3}$$

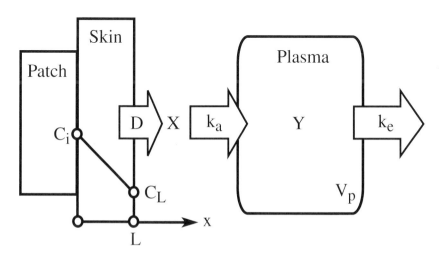

FIGURE 10.2 Scheme of the process with the patch, the skin, and the blood compartment.

Under these conditions, the skin plays the role of a membrane for the drug transport, with constant concentrations on each surface. The drug concentration developed through the thickness L of the skin is thus expressed by the following series [9,10]:

$$\frac{C_{x,t}}{C_0} = 1 - \frac{x}{L} - \frac{2}{\pi}\sum_{n=1}^{\infty}\frac{1}{n}\sin\frac{n\pi x}{L}\exp\left(-n^2\pi^2\frac{Dt}{L^2}\right) \tag{10.4}$$

The amount of drug that emerges from the internal face of the skin is

$$\frac{M_t}{C_0} = \frac{Dt}{L} - \frac{L}{6} - \frac{2L}{\pi^2}\sum_{n=1}^{\infty}\frac{(-1)^n}{n^2}\exp\left(-n^2\pi^2\frac{Dt}{L^2}\right) \tag{10.5}$$

Equation 10.4 shows that the concentration of drug in the skin $C_{x,t}$ increases with time, and the profiles of drug concentration through the thickness of the skin tend to become linear when the series vanishes.

In the same way, from Equation 10.5, the amount of drug leaving the skin increases slowly with time as the series in this equation vanishes with time. Thus, after a long period of time Equation 10.5 tends to become linear, and the expression of the asymptote is

$$M_t = \frac{DC_0}{L}\left(t - \frac{L^2}{6D}\right) \tag{10.6}$$

the slope of which is

$$Slope = \frac{M_t}{t} = \frac{DC_0}{L} \tag{10.7}$$

and the intercept on the time-axis is

$$t_i = \frac{L^2}{6D} \tag{10.8}$$

10.3.2.1   Transfer into the Blood Compartment

The amount of drug at the internal surface of the skin, X_t, is given by the following relation:

$$\frac{dX}{dt} = \frac{dM}{dt} - k_a X_t \tag{10.9}$$

The amount of drug in the blood compartment is obtained by the relationship:

$$\frac{dY}{dt} = k_a X - k_e Y \tag{10.10}$$

where k_a and k_e are the rate constants of the first-order kinetics of absorption and elimination.

Of course, the amount of drug eliminated is also obtained, by the relation:

$$\frac{dZ}{dt} = k_e Y \tag{10.11}$$

Calculation is made step by step, using a numerical method with a constant increment of time, because no analytical solution of the problem can be found.

10.4 RESULTS FOR THE TRANSDERMAL DELIVERY OF METOPROLOL

10.4.1 EXPERIMENTAL PART [11]

Metoprolol, a β_1-selective adrenergic blocking agent is a first-choice drug in the treatment of mild to moderate hypertension and stable angina and is beneficial in postinfarction patients. But it is also subjected to extensive hepatic first-pass metabolism following oral administration and has a short biological half-life [12]. Thus, the transdermal route of administration can avoid the hepatic first-pass effect and achieve higher systemic bioavailability of this drug. In this paper, nicely reported, various experiments have been made, such as the preparation of the multilaminate adhesive device able to deliver the drug transdermally, the skin permeation studies on Valia–Chien glass cell diffusion, whole bioavailability studies with the plasma level determination of the drug in hairless rats after an i.v. bolus injection and the plasma level obtained in the same rats after oral administration, and finally the determination of the plasma levels in these rats after application of an adhesive device [11].

From these experiments, it is possible to determine the values necessary to calculate the plasma drug level (e.g., the characteristics of the skin of the rats, such as its thickness; the diffusivity while the concentration of the drug on the skin surface is calculated; and the pharmacokinetic parameters of the rats, such as the plasmatic volume and the rate constants of absorption and elimination).

10.4.1.1 Preparation of the Patch [11]

A weighed amount of metoprolol base was mixed thoroughly with the specially formulated polyacrylate adhesive, and a uniform layer of a fixed thickness (1.2 mm) was made on a heat-sealable backing membrane with a laboratory coating device. The whole system was cured overnight at room temperature under a hood in a dust-free environment. The laminate was then covered by a release liner, cut into 10 cm² pieces, and used in the subsequent experiments.

10.4.1.2 Skin Permeation Studies [11]

The freshly excised full-thickness hairless rat skin was mounted on Valia–Chien glass diffusion cells, with the stratum corneum side in intimate contact with the drug-releasing surface of the patch and the dermal side facing the receptor solution. This solution (pH 7.4 Sorensen's phosphate buffer), kept at $37°C$ and stirred conveniently, was analyzed at intervals by withdrawing samples and replacing them by an equal volume of fresh solution. The skin condition was verified in the glass diffusion cells.

10.4.1.3 Bioavailability Studies [11]

Three categories of rats were considered for the following three series of experiments:

- In the first group (6), the rats received an i.v. administration of drug (50 mg/kg) and blood samples were withdrawn at intervals for analysis.
- The second group of rats (6) was administered a single 75 mg/kg oral dose, and blood samples were collected at intervals for analysis.
- The third group (6) received transdermal administration on the abdomen with a 10 cm² adhesive device containing the drug dose, and blood samples were taken for analysis. The transdermal dose and rate of delivery were calculated from the in vitro hairless rat skin permeation studies.

10.4.2 RESULTS OBTAINED FROM IN VITRO EXPERIMENTS

The first part of the results is concerned with the determination of the parameters of interest to the drug transfer. The main drug transfer of interest is concerned with the patch and the diffusion through the skin. The kinetics of drug transfer through the patch–skin system is shown in Figure 10.3, obtained either from experiments (dotted line) [11] or from calculation (solid line).

The following comments can be drawn from the kinetics calculated using Equation 10.5:

- The kinetics of drug transfer through the patch–skin system corresponds with the transfer through a membrane with constant concentration on each side [9, 10].
- From the beginning to about 5 to 7 h into the process, the amount of drug transferred increases slowly with time, following Equation 10.5, and the process is controlled by transient diffusion, meaning that the gradient of concentration through the thickness of the skin is not yet linear.
- After 5 to 7 h, the kinetics becomes linear, the process being controlled by stationary diffusion, as the series in Equation 10.5 vanishes and the gradient of concentration through the thickness of the skin is constant with time and linear.
- From the intercept on the time-axis, it is possible to obtain a relationship between the thickness of the skin and the diffusivity of the drug, according to Equation 10.8.

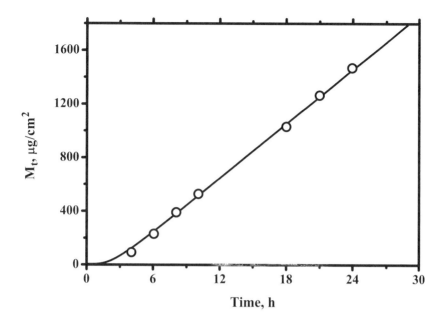

FIGURE 10.3 Kinetics of drug transfer (metoprolol) through the skin: calculated using Equation 10.5 (solid line); experimental in-vitro data on hairless rats (dotted line). Reprinted From Polymer Testing pp. 889–897, 2000. With permission from Elsevier.

- The value of the slope, in the linear part of the kinetics of drug transport through the skin, gives a relationship between the same parameters shown in Equation 10.7. In addition, the drug concentration C_0 is maintained on the external surface of the skin by the patch.
- As the thickness of the skin has been measured experimentally [11], the values of the other parameters can easily be evaluated using Equation 10.7 and Equation 10.8.
- The good agreement between the calculated and experimental curves in Figure 10.3 proves the accuracy of the values of the diffusion parameters given in Table 10.1.
- It is worth noting that the patch is capable of maintaining the drug available at a constant concentration C_0 on the surface of the skin over a period of at least 24 h, as illustrated by the shape of the curve in Figure 10.3.

TABLE 10.1
Parameters of Diffusion through the Skin

Slope = 68×10^{-6} g/cm^{2}hour;	t_i = 2.6 h;	L = 0.035 cm;	$D = 2.2 \times 10^{-8}$ cm^2/s;
C_0 = 0.03g/cm^3;	$h = 6 \times 10^{-6}$ cm/s.		

10.4.3 Results Concerning In Vivo Experiments

The pharmacokinetic parameters are obtained from the drug level–time history obtained in the blood [11]. From the i.v. bolus injection, the rate constant of elimination k_e is related to the time of half-life of the drug:

$$Ln2 = 0.693 = k_e t_{0.5}$$

From the oral absorption of the drug [11], it is possible to determine the rate constant of absorption either by calculation, using the main characteristics of the peak (see Equation 1.6 for the time at which the peak appears), or by using a numerical model using the whole plasma drug curve.

The volume of distribution is evaluated [11] from the plasma drug level using Equation 1.5.

The values of the pharmacokinetic parameters are shown in Table 10.2.

10.4.4 Evaluation of the Plasma Drug Level

Using the numerical model described in the theoretical part of this chapter (Section 10.3.2) and the data shown in Table 10.1 and Table 10.2, the plasma drug level obtained by transdermal delivery has been calculated. In Figure 10.4, the plasma drug profiles obtained either by calculation (solid line) [8] or by experiments (dotted line) [11] are drawn, leading to some conclusions:

- Rather good agreement can be observed between the experimental curve and the calculated curve. The mean values of the experiments are shown, as well as the upper and lower values obtained with six rats corresponding to their intervariability.
- Just as for the kinetics of the drug transfer through the skin, it takes some time for the drug to get into the blood compartment. For up to 5 to 7 h, under transient diffusion through the skin, the plasma drug level increases.
- After around 10 h, stationary conditions are attained, and the plasma drug level remains constant up to the removal of the patch.
- When the patch is removed after 24 h of drug delivery, the plasma drug level decreases very quickly, as shown by curve 2 in Figure 10.4. A problem arises in calculating the drug level using the numerical model after the removal of the patch. Two extreme assumptions can be made for the drug located in the skin:

TABLE 10.2
Pharmacokinetic Parameters
of Metoprolol in Rats [11]

$k_a = 1.1/h$	$k_e = 0.78/h$	$V_p = 91$

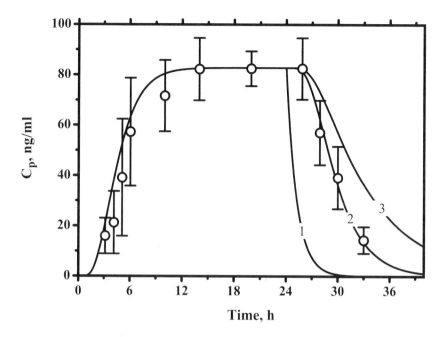

FIGURE 10.4 Plasma drug level (metoprolol) obtained with the patch on hairless rats: calculated using the numerical model (solid line); experimental in vivo data (dotted line). Curve 1 with $k_a = 0$; curve 2 with $h = 6.10^{-6}$ cm/s; curve 3 with $h = 0$. Reprinted From Polymer Testing pp. 889–897, 2000. With permission from Elsevier.

- This drug remains in the skin, and calculation is made by writing that as $k_a = 0$.
- This drug enters the blood compartment entirely by diffusion, resulting from the gradient of concentration. Thus calculation is made by keeping k_a constant and $\frac{\partial C}{\partial x} = 0$ at the external surface, meaning that the drug located in the skin is diffusing only and totally into the blood compartment.
- Because neither curve 1 (obtained with the first assumption) nor curve 3 (determined using the second assumption) is able to fit the experimental curve expressing the decrease in drug concentration after the patch has been removed, another assumption is made, by considering that the drug can leave the skin on both sides. Thus the drug transfer into the blood compartment is maintained by keeping the rate constant of absorption constant and writing that there is a drug transfer outside the skin obeying the following equation:

$$-D\frac{\partial C}{\partial x} = hC \tag{10.12}$$

Equation 10.12 expresses that the rate at which the drug is brought to the external surface of the skin by diffusion is constantly equal to the rate of elimination outside the skin, h being the coefficient of the transfer of the drug out of the skin. The value of h given in Table 10.1 enables the theoretical curve to fit the experimental one. It can be said that such a coefficient of convection exists when a substance diffuses out of a solid [9,10], and it plays an important role in the transport of a contaminant or food at the package–food interface [13].

10.4.5 CONCLUSIONS AND FUTURE PROSPECTS

Various studies have been carried out on transdermal drug delivery, and even a large bibliography consisting of 148 references [2] cannot be exhaustive. Some other papers on the subject of diffusion are also worth noting.

In the first, diffusion phenomena have been deeply explored not only through the skin but also in the tissue between the skin and the capillaries [14]. As a result, the drug concentration profiles developed through the dermis have been determined experimentally. The remaining problem is that the calculated plasma drug level did not correspond with the experimental one, especially at the beginning of the process; this model predicted that the steady state would be reached after around 35 to 50 min, whereas the experimental values for this time are much larger.

The effect of patch placement site has been studied by making in vitro experiments, showing that the upper dorsal position allows a larger rate of drug delivery for humans and rats [15].

The effect of enhancers has also been widely studied, and only a few reports are mentioned here. There is the relative effect of laurocapram (Azone) and transcutol on the solubility and diffusivity of a drug, 4-cyanophenol, in human stratum corneum [16]. A deep mechanistic investigation of the in vitro human skin permeation–enhancing effect of laurocapram was made, leading to the conclusion that the lower the lipophilicity value of the drug, the greater the enhancer effect observed [17]. Other complex models were used for predicting the penetration of monofunctional solutes from aqueous solutions through the human stratum corneum [18].

The contribution of the various skin layers to the resistance to permeation has been considered, showing not only that the stratum corneum is the main barrier to permeation but also that there is a threefold difference between the fluxes through unstripped and stripped full-thickness for a drug [19].

The effect of the adhesive layer, the part of the patch in contact with the skin, has been deeply studied [4], showing that the drug concentration should reach saturation in this polymeric layer, in order not to play the role of a barrier.

Disposition of a multiple-dose transdermal system was considered with Nicotine TTS [20], and the analysis of the components of variance contributing to the variability in nicotine delivery indicated that the in vivo transdermal permeation of that drug is rate limited by both the device and the intrinsic skin diffusivity.

Various studies have explored the possibility of building in vitro–in vivo correlations in the same way as those managed by the FDA for oral dosage forms [21]. These studies conclude that there are some difficulties in accomplishing this task.

If the first-pass hepatic is generally circumvented with transdermal drug delivery, a possibility of metabolism remains through the skin, so propanolol is partly metabolized in human skin tissues and its metabolites retained in the skin [22, 23].

The aging effect is also of great importance in the transdermal transfer, because the permeation properties of the skin alter with age. If the absorption rate does not change very much for lipophilic materials, such as testosterone or estradiol, this rate decreases significantly for less lipophilic materials, such as hydrocortisone or acetyl salicylic acid. This is due to various factors: The time necessary for repairing the stratum corneum increases and the number of subdermis capillaries decreases with age [24, 25].

Comparisons of the pharmacokinetics of transdermal systems have been made. The first clinical study on humans compared the pharmacokinetic parameters and the plasma concentration time profiles of the nicotine transdermal products Nicoderm and Habitrol [26], in either single or multiple doses. Thus, the observed differences in plasma profiles were attributed to the technical designs of these two systems. The nicotine concentration present in the adhesive layer acts on the rapid rise in the plasma concentration, whereas the rate-controlled membrane determining a constant rate of drug from the reservoir is responsible for a steady plasma nicotine concentration [26].

Conventional patch-based transdermal systems are easy to use, but they are not able to deliver drugs with a large molecular weight, such as peptides or proteins. Thus, various improvements have been developed [2]. A brief survey is given here:

- Another way to increase the permeability of the skin consists of using the transdermal iontophoresis by applying a potential electrical gradient between the skin site and another place on the body with two electrodes. The electrode on the skin bears the same charge as the drug ion, and the movement of these ions through the skin is brought about using an appropriate direct–alternating current between the two electrodes. Of course, this technique is useful for drugs that are predominantly ionized [27–30]. Nevertheless, much faster transport of ions facilitates the transport of neutral molecules, because the electric current increases the permeability of the stratum corneum. It must also be noted that some interactions may occur between penetration enhancers and iontophoresis [31].
- Electroporation, in making the stratum corneum porous by way of electric discharges, greatly increases the rate of drug transport through the skin.
- Transdermic administration, using ultrasound, is called *sonophoresis* [32].
- A patch containing microneedles able to perforate the stratum corneum and project the drug in the dermis can be used. This system can work either in bolus or in a continuous way.

10.5 EFFECT OF THE CHARACTERISTICS OF THE TTS

On the whole, the TTSs can be classified into three categories:

- The first, very simple, delivers the drug to the skin by diffusion through the polymer layers of the system.
- Another system, more complex, consists of these layers and a reservoir for the drug.

- The third system differs from the other two in the sense that the drug is transferred not by diffusion but by convection to the external surface of the skin.

The objective of this section is to evaluate the ability of each of these three systems to deliver the drug, by calculating either the rate of drug diffusing through the skin or the plasma drug level. Of course, the drug profiles developed through the TTS itself are also calculated and drawn, in order to make the whole process understandable. The scheme of these three systems is drawn in Figure 10.5.

10.5.1 DIFFUSION-CONTROLLED TTS WITH A POLYMER ALONE

In this monolithic system, the TTS is generally made of three layers (Figure 10.5, scheme 1):

- An impermeable backing
- An intermediate polymer matrix containing the drug
- The skin adhesive layer, which is attached to a release liner before use

In this type of system, the matrix polymer is designed to control the diffusion of the drug through and from the system into the skin. Initially the drug is uniformly distributed throughout the polymer matrix. When the system is placed on the skin, the drug permeates the skin, provoking a decrease in the drug concentration on the

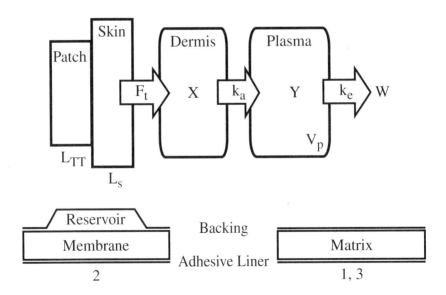

FIGURE 10.5 Scheme of the three types of TTS. 1. Simple device with the backing, the polymer matrix, and the adhesive liner. 2. The device also comporting a drug reservoir. 3. The device similar to that shown in (1), with the difference that the drug transport is controlled by convection through the polymeric matrix.

surface in contact with the skin. It stands to reason that the value of the diffusivity of the drug through the polymer, as compared to that in the skin, plays a major role in the rate of drug transport through the skin and on the plasma drug level.

10.5.1.1 Assumptions

The following assumptions can describe the process and thus are used for calculation:

- The drug concentration is initially uniform in the polymer matrix.
- The TTS and the skin are in perfect contact.
- Initially, the skin is free of drug, at least for the first dose.
- In order to facilitate comparison, the drug concentration is the same for the three types of TTS. Moreover, the concentration is initially the same in the three TTSs.
- The process of drug transfer is controlled by diffusion, in either the TTS or the skin, with a constant diffusivity in both these media.
- The drug concentration on the side of the skin opposite to the TTS is very low, because the drug is absorbed in the plasma with a first-order kinetics.

The theoretical treatment is made by using the numerical model described [3].

The results are expressed in three ways, with the profiles of drug concentration developed through either the TTS or the skin, and with the kinetics of drug transfer into the dermis, as well as with the plasma drug level obtained under these conditions.

The profiles of drug concentration are drawn in the TTS–skin system (Figure 10.6) when the diffusivity is much larger in the TTS than in the skin (e.g., when the diffusivity in the TTS is 200 times that in the skin).

The kinetics of drug transfer into the dermis are drawn for different values of the diffusivity in the TTS (Figure 10.7), being successively equal to that in the skin (curve 4), then 20 times that in the skin (curve 3), 200 times larger (curve 2), and finally 2,000 times that in the skin (curve 1).

The four plasma drug profiles obtained under these for conditions are drawn in Figure 10.8.

In Figure 10.7 and Figure 10.8, the values obtained for metoprolol in the rats are also shown in comparison [8,11].

From these curves, the following observations can be made:

- The profiles of drug concentration developed through the TTS–skin system (Figure 10.6) afford further insight into the nature of the process of drug transfer controlled by diffusion. The drug concentration progresses continuously through the skin under transient conditions. After a time of about 1 h, the drug reaches the dermis. After a time of more than 6 h, the drug concentration becomes nearly uniform in the TTS and tends to be linear through the skin. The drug concentration decreases regularly with time in the TTS and on the skin surface in contact with the TTS.
- The kinetics of drug transfer in the dermis exhibits a typical pattern that follows the drug concentration profiles developed through the

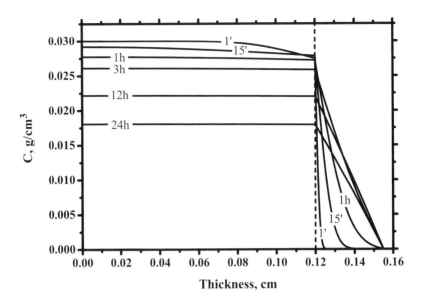

FIGURE 10.6 Profiles of drug concentration developed through the TTS and skin, with the polymer alone and the drug transport controlled by diffusion. Initial drug concentration: 0.03 g/ml. Thickness of the patch: 0.12 cm, of the skin: 0.035 cm. Diffusivity in the TTS: 4.4×10^{-6} cm^2/s, and in the skin: 2.2×10^{-8} cm^2/s. The times are noted.

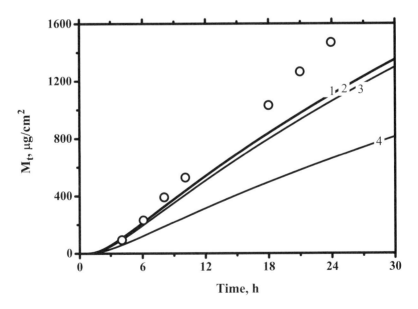

FIGURE 10.7 Kinetics of drug transfer through the skin, for various values of the diffusivity in the polymeric device. 1. $D = 4.4 \times 10^{-5}$ cm^2/s. 2. $D = 4.4 \times 10^{-6}$ cm^2/s. 3. $D = 4.4 \times 10^{-7}$ cm^2/s. 4. $D = 2.2 \times 10^{-8}$ cm^2/s. Skin $D = 2.2 \times 10^{-8}$ cm^2/s, $L = 0.035$ cm. Reprinted From Computational Theoretical Polymer Sci. 10, pp. 391–401, 2000. With permission from Elsevier.

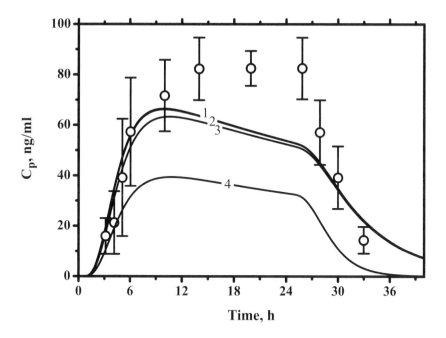

FIGURE 10.8 Plasma drug level obtained with the polymeric device and various values of the diffusivity. Calculated (solid line) with: 1. $D = 4.4 \times 10^{-5} cm^2/s$; 2. $D = 4.4 \times 10^{-6} cm^2/s$; 3. $D = 4.4 \times 10^{-7} cm^2/s$; 4. $D = 2.2 \times 10^{-8} cm^2/s$. Drug: metoprolol with the pharmacokinetic data shown in Table 10.2.

TTS–skin system. Thus, with the larger value of the diffusivity, curve 1 in Figure 10.7 follows that obtained in experiments [11] only at the beginning of the process, when the concentration in the TTS and on the skin surface has not decreased so much below the value of the concentration shown in Table 10.1, which is associated with the experimental curve (dotted line). When the concentration in the TTS has decreased significantly below this concentration, the slope in Figure 10.7, which represents the rate of drug transfer through the skin, continuously decreases. Thus the obvious statement holds: the higher the diffusivity in the TTS, the higher the kinetics of drug delivery in the dermis.

- Up to 3 h, the kinetics of transfer determined by calculation (curve 1) follows that obtained by experiment. After this time interval, when the drug concentration on the TTS surface in contact with the skin decreases significantly, the plasma drug levels obtained by calculation are below those obtained from experiment. Thus, the effect of the diffusivity in the TTS on the plasma drug profile is a consequence of the results shown in Figure 10.8.
- After removal of the TTS from the skin, the plasma drug level decreases, resulting from the kinetics of elimination. Only a part of the drug located in the skin is delivered into the plasma, the other part leaving the skin according to Equation 10.12.

10.5.2 DIFFUSION-CONTROLLED TTS WITH A RESERVOIR

10.5.2.1 Assumptions

The assumptions shown in Section 10.5.1 still hold, except one, resulting from the presence of the reservoir that maintains the drug concentration constant on the external surface of the TTS opposite to the surface in contact with the skin: *The drug concentration is constant on the side of the TTS opposite to the skin.*

The process of drug transfer is to some extent similar to that shown without a reservoir (Section 10.5.1), the main difference being that the reservoir maintains a constant concentration of drug on the surface of the polymer in the TTS.

The results are expressed in terms of the profiles of concentration attained under stationary conditions in the TTS–skin system (Figure 10.9) and the kinetics of drug transfer through the skin (Figure 10.10), as well as the plasma drug level obtained in various cases (Figure 10.11).

The profiles of concentration attained under the nearly stationary conditions through the TTS–skin thickness are drawn (Figure 10.9) for various values of the diffusivity in the TTS and for a constant drug concentration ($C = 0.03$ g/ml) on the external surface of the TTS. The kinetics of drug transfer through the skin are also drawn (Figure 10.10) for the same values of the diffusivity in the TTS and the experimental values shown in Table 10.2 [8,11].

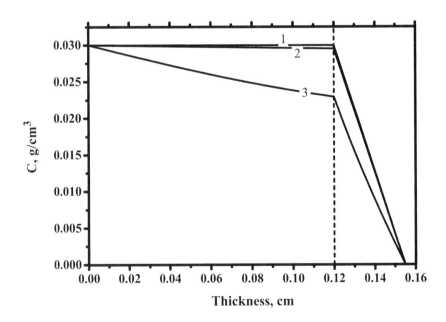

FIGURE 10.9 Profiles of drug concentration through the TTS and skin under stationary conditions, with a polymer device and a reservoir, for various values of the diffusivity of the drug in the TTS polymer. Skin: $L = 0.035$ cm; $D = 2.2 \times 10^{-8}$ cm^2/s; drug concentration = 0.03 g/ml. D in the TTS: 1. 4.4×10^{-5} cm^2/s; 2. 4.4×10^{-6} cm^2/s; 3. 4.4×10^{-7} cm^2/s.

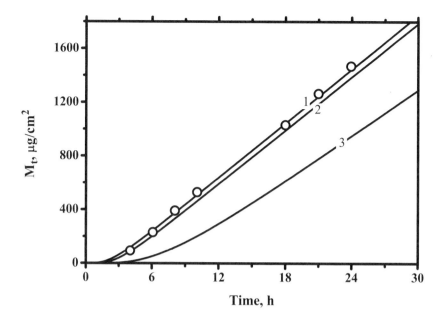

FIGURE 10.10 Kinetics of drug transfer through the skin, with a TTS with a reservoir, for various values of the diffusivity in the TTS polymer. Drug: metoprolol. Skin: $L = 0.035$ cm; $D = 2.2 \times 10^{-8}$ cm^2/s; drug concentration = 0.03 g/ml; D in the TTS: 1. 4.4×10^{-5} cm^2/s; 2. 4.4×10^{-6} cm^2/s; 3. 4.4×10^{-7} cm^2/s. Reprinted From Computational Theoretical Polymer Sci. 10, pp. 391–401, 2000. With permission from Elsevier.

The plasma levels are drawn in Figure 10.11, as they are calculated for the various values of the diffusivity of the drug in the TTS.

A few comments are worth noting from these curves:

- Of course, the linear gradients of drug concentration are inversely proportional to the values of the diffusivity of the drug in the TTS. In other words, the drug concentration remains nearly uniform in the TTS when the diffusivity is very high, as in curve 1.
- For lower diffusivity values, the drug concentration does not remain uniform in the TTS; moreover, the drug concentration on the skin surface decreases during the process.
- The kinetics of drug transfer through the skin calculated with the two highest diffusivity values are in good agreement with the experimental kinetics (Figure 10.10, curves 1 and 2). But when the diffusivity of the drug is too low in the TTS, the concentration on the skin surface decreases, and the kinetics of drug transfer through the skin is lower than the experimental kinetics (Figure 10.10, curve 3).
- The plasma drug levels calculated with the two higher diffusivity values in the TTS are nearly similar to that obtained by experiment, from the beginning to the end of the process, as shown in Figure 10.11 (curves 1 and 2).

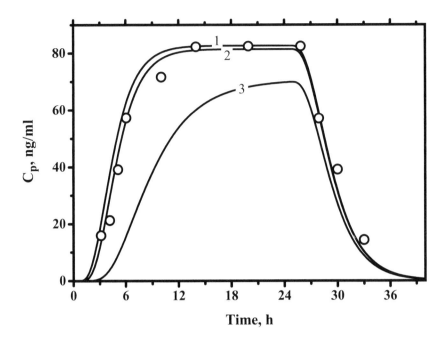

FIGURE 10.11 Plasma drug levels obtained with the TTS and reservoir for various values of the diffusivity in the TTS polymer. Drug: metoprolol. Skin: $L = 0.035$ cm; $D = 2.2 \times 10^{-8}$ cm^2/s; drug concentration $= 0.03$ g/ml; D in the TTS: 1. 4.4×10^{-5} cm^2/s; 2. 4.4×10^{-6} cm^2/s; 3. 4.4×10^{-7} cm^2/s.

- With the lower diffusivity value of the drug in the TTS (Figure 10.11, curve 3), the plasma drug level remains similar to that obtained from experiment from the beginning to about 3 h; after this time, the plasma drug level increases more slowly and reaches a plateau at a lower level than that obtained from experiment.
- Just as for the monolithic device, after removal of the TTS from the skin, the plasma drug level sharply decreases. The coefficient of transfer at the external side of the skin used for calculation is the same as that used for the monolithic device.

10.5.3 TTS with Drug Release Controlled by Convection

The main effect of convection is that the rate of drug transfer through the TTS is so high that the drug concentration on the skin surface is maintained as constant, at least for a time period over which the mean drug concentration remains nearly constant in the TTS.

The results are expressed by the drug concentration profiles developed through the TTS–skin system (Figure 10.12), the kinetics of drug transfer through the skin (Figure 10.13), and the plasma drug profiles (10.14).

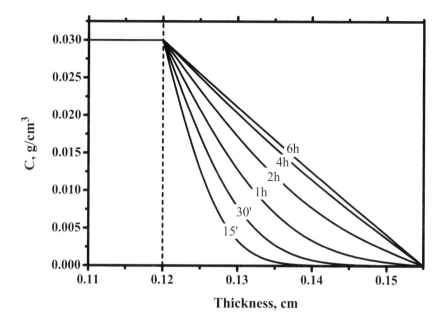

FIGURE 10.12 Profiles of drug concentration through the TTS and skin with TTS with drug convective transfer, under transient and stationary conditions. Drug: metoprolol. Skin: $L = 0.035$ cm; $D = 2.2 \times 10^{-8}$ cm^2/s; drug concentration $= 0.03$ g/ml.

These curves lead to the following conclusions:

- The main conclusion is that the drug transfer through the TTS is so large that the concentration of the drug is maintained as constant on the skin surface. The usual concentration profiles develop through the skin thickness under transient conditions; after a time depending on the diffusivity of the drug through the skin and its thickness, the stationary state is attained with a linear profile of the concentration through the skin (Figure 10.12).
- The kinetics of drug transfer through the skin are drawn for various values of the constant drug concentration in the TTS ranging from 0.075 to 0.01 g/ml (Figure 10.13). Of course, the higher the drug concentration in the TTS, the higher the rate of drug transfer through the skin, provided that the solubility of the drug in the skin is sufficiently high. In fact, when the process has reached stationary conditions, the rate of drug transfer through the skin is proportional to this drug concentration. Note that when the solubility of the drug in the skin is not so large, an increase in the concentration in the TTS will find a limit for the concentration in the skin, and thus a partition factor will occur; nevertheless, the concentration of the drug in the skin is limited by the solubility of the drug in the skin.
- A good agreement is observed between the theoretical and experimental drug levels when the constant concentration of the drug is 0.03 g/ml on the

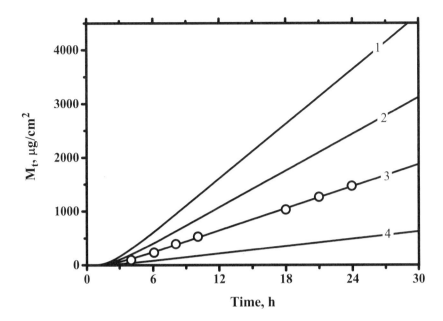

FIGURE 10.13 Kinetics of drug transfer through the skin with the TTS with drug convective transfer, and various values of the constant drug concentration in the TTS. Skin: $L = 0.035$ cm; $D = 2.2 \times 10^{-8}$ cm^2/s; drug concentration: 1. 0.075 g/ml; 2. 0.05 g/ml; 3. 0.03 g/ml; 4. 0.01 g/ml. Reprinted From Computational Theoretical Polymer 10, pp. 391–401, 2000. With permission From Elsevier.

skin surface (Figure 10.14, curve 3). As shown with the other curves, the plasma drug level depends largely on the value of this constant concentration of the drug on the surface of the skin. In fact, the drug level at the plateau is proportional to the drug concentration in the TTS applied to the skin surface or, rather, to the drug concentration on the skin surface; this concentration on the skin surface cannot be higher than the solubility of the drug in the stratum corneum and the skin.

- The time necessary for the drug to cross the skin before reaching the bloodstream is the same whatever the drug concentration in the skin; in fact, this lag time depends only on the characteristics of the skin (e.g., its thickness and the drug diffusivity through it).
- After removal of the TTS, the plasma drug level falls in all cases, and Equation 10.12 applies, meaning that only a part of the drug located in the skin is delivered to the plasma.

10.6 EFFECT OF THE SKIN PARAMETERS

For the transdermal drug delivery, the effect of the characteristics of the skin on the plasma drug level is concentrated in a few lines, giving a theoretical basis for the attempts made to increase the efficacy of the TTS.

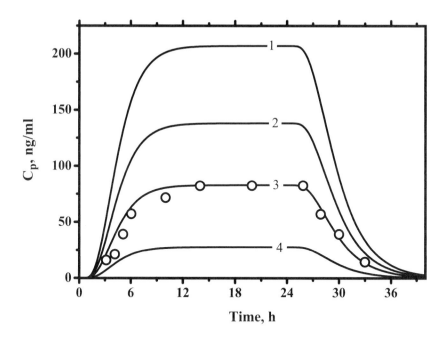

FIGURE 10.14 Plasma drug levels with the convective TTS and various values of the constant drug concentration in the TTS and the surface of the skin. Metoprolol with $k_a = 1.1/h$ and $k_c = 0.78/h$. Skin: $L = 0.035$ cm; $D = 2.2 \times 10^{-8}$ cm^2/s; drug concentration: 1. 0.075 g/ml; 2. 0.05 g/ml; 3. 0.03 g/ml; 4. 0.01 g/ml. Reprinted From Computational Theoretical Polymer Sci. 10, pp. 391–401, 2000. With permission from Elsevier.

Two methods should be considered:

- Increasing the diffusivity of the drug through the skin
- Increasing the drug concentration within the skin

10.6.1 INCREASE IN THE SOLUBILITY OF THE DRUG IN THE SKIN

As shown in Equation 10.7, the rate of drug transfer through the skin is proportional to the concentration of the drug in the skin; in the same way, an increase in the drug concentration in the skin increases the plasma drug level. This result is displayed in Figure 10.14, where the various values of the plasma drug level calculated for different values of the concentration of the drug on the surface of the skin show that the plasma drug level is proportional to the drug concentration in the skin. Let us remark again that the drug concentration in the skin is limited by its solubility. Thus, any method that can increase the solubility of the drug in the stratum corneum is of great interest, such as the use of enhancers or other techniques described in reviews [1, 2].

10.6.2 INCREASE IN THE DIFFUSIVITY OF THE DRUG IN THE SKIN

By considering Equation 10.6 expressing the rate of drug transfer through the skin, it clearly appears that this rate is proportional to the diffusivity; moreover, the lag

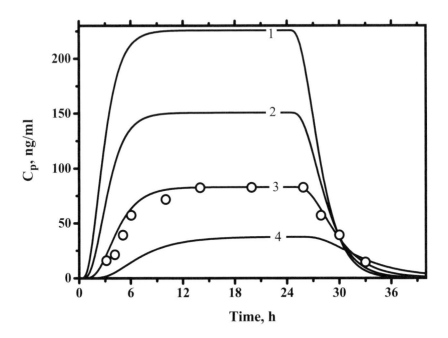

FIGURE 10.15 Plasma drug level with the convective TTS and various values of the diffusivity of the drug in the skin. Metoprolol with $k_a = 1.1/h$ and $k_e = 0.78/h$. Skin: $L = 0.035$ cm; drug concentration on the skin surface: 0.03 g/ml. $D \times 10^{-8}$ cm^2/s: 1. 6; 2. 4; 3. 2.2; 4. 1.

time necessary for the drug to cross the skin is inversely proportional to this diffusivity. For a given drug characterized by its rate constants of absorption and of elimination, the amount of drug at the internal surface of the skin increases with this diffusivity. The plasma drug level also increases when this diffusivity is increased, as shown in Figure 10.15. This remark is an argument in favor of any method able to increase the diffusivity, such as those based on the techniques of iontophoresis, high-frequency ultrasound, or microneedles.

REFERENCES

1. Heilmann, K., *Therapeutic Systems,* Georg Thieme Verlag, Stuttgart, 37, 1984.
2. Falson-Rieg, F. and Pirot, F., Dispositifs transdermiques, *Nouvelles Formes Médicamenteuses,* Ed. Tec. et Doc., Paris, 14, 259, 2004.
3. Ouriemchi, E.M. and Vergnaud, J.M., Processes of drug transfer with three different polymeric systems with transdermal drug delivery, *Comput. Theoret. Polymer Sci.,* 10, 391, 2000.
4. Roy, S.D. et al., Controlled transdermal delivery of fentanyl: characterizations of pressure-sensitive adhesives for matrix patch design, *J. Pharm. Sci.,* 85, 491, 1996.
5. Ouriemchi, E.M., Ghosh, T.P., and Vergnaud, J.M., Modelling the transdermal transfer of metoprolol, Am. Chem. Meet., Boston, August 23–27, 1998.

6. Ouriemchi, E.M., Ghosh, T.P., and Vergnaud, J.M., Assessment of plasma drug level with transdermal delivery, Ame Chem. Meet., Div. Medicinal Chem., Anaheim, CA, March, 1999.

7. Ghosh, T.P., Rosca, I.D., and Vergnaud, J.M., Process of transdermal drug transfer from a polymer device, 3rd World Meeting on Pharmaceutics, Berlin, April 2000.

8. Ouriemchi, E.M., Ghosh, T.P., and Vergnaud, J.M., Transdermal drug transfer from a polymer device: study of the polymer and the process, *Polym. Testing,* 19, 889, 2000.

9. Crank, J., *The Mathematics of Diffusion,* Clarendon Press, Oxford, 4.3.3, 49, 1975.

10. Vergnaud, J.M., *Controlled Drug Release of Oral Dosage Forms,* Ellis Horwood, Chichester, U.K., 2.2.3, 35, 1993.

11. Ghosh, T.P. et al., Transdermal delivery of Metropolol: in vitro skin permeation and bioavailability in hairless rats, *J. Pharm. Sci.,* 84, 158, 1995.

12. Regardh, C.G. and Johnsson,G., Clinical pharmacokinetics of metropolol, *Clin. Pharmacokin.,* 5, 557, 1980.

13. Laoubi, S. and Vergnaud, J.M., Processes of chemical transfer from a packaging into liquid or solid food by diffusion-convection or by diffusion, *Plastics, Rubber Compos. Process. Applic.,* 25, 83, 1996.

14. Gao, K., Wientjes, M.G., and Au, J.L., Use of drug kinetics in dermis to predict in vivo blood concentration after topical application, *Pharm. Res.,* 12, 2012, 1995.

15. Ho, H. and Chien, Y.W., Kinetic evaluation of transdermal nicotine delivery systems, *Drug Devel. Ind. Pharm.,* 19, 295, 1993.

16. Harrison, J.E. et al., The relative effect of Azone and Transcutol on permeant diffusivity and solubility in human stratum corneum, *Pharm. Res.,* 13, 542, 1996.

17. Diez-Sales, O. et al., A mechanistic investigation of the in vitro human skin permeation enhancing effect of Azone, *Int. J. Pharm.,* 129, 33, 1996.

18. Roberts, M.S. et al., Epidermal permeability-penetrant structure relationships: 1. An analysis of methods of predicting penetration of mono-functional solutes from aqueous solutions, *Int. J. Pharm.,* 126, 219, 1995.

19. Anigbogu, A.N.C., Williams, A.C., and Barry, B.W., Permeation characteristics of 8-Methoxypsoralen through human skin; relevance to clinical treatment, *J. Pharm. Pharmacol.,* 48, 357, 1996.

20. Kochak, G.M. et al., Pharmacokinetic disposition of multiple-dose transdermal nicotine in healthy adult smokers, *Pharm. Res.,* 9, 1451, 1992.

21. Hadgraft, J., Pharmaceutical aspects of transdermal nitroglycerin, *Int. J. Pharm.,* 135, 1, 1996.

22. Krishna, R. and Pandit, J.K., Carboxymethyl cellulose-sodium based transdermal drug delivery system for propanolol, *J. Pharm. Pharmacol.,* 48, 367, 1996.

23. Ademola, J.I. et al., Metabolism of propanolol during percutaneous absorption in human skin, *J. Pharm. Sci.,* 82, 767, 1993.

24. Mary, S., Makki, S., and Agache, P., Perméabilité cutanée en fonction de l'âge, *Cosmétologie,* 12, 34, 1996.

25. Harwell, J.D. and Maibach, H.I., Percutaneous absorption and inflammation in aged skin: a review, *J. Am. Acad. Dermatol.,* 31, 1015, 1994.

26. Gupta, S.K. et al., Comparison of the pharmacokinetics of two nicotine transdermal systems: Nicoderm and Habitrol, *J. Clin. Pharmacol.,* 35, 493, 1995.

27. Green, P.G. et al., Transdermal iontophoresis of amino acids and peptides in vivo, *J. Control. Release,* 21, 187, 1992.

28. Brand, R.M. and Guy, R.H., Iontophoresis of nicotine in vitro: pulsatile drug delivery across the skin, *J. Control. Release,* 33, 285, 1995.

29. Hirvonen, J., Kalia, Y.N., and Guy, R.H., Transdermal delivery of peptides by ionto-phoresis: structure, mechanism and feasibility, *Nature Biotech.,* 14, 1710, 1996.
30. Pirot, F. et al., Stratum corneum thickness and apparent water diffusivity: facile and non invasive quantitation in vivo, *Pharm. Res.,* 15, 490, 1998.
31. Kalia, Y.N. and Guy, R.H., Interaction between penetration enhancers and iontophore-sis, *J. Control. Release,* 44, 33, 1997.
32. Bommannan, D. et al., Sonophoresis: the use of high-frequency ultrasound to enhance transdermal drug delivery, *Pharm. Res.,* 9, 559, 1992.

Conclusions

Very often, a book's conclusions are a summary of the answers to the main questions set in the Introduction or the Foreword. In this case, all the answers to these queries have been largely described in the text. Nevertheless, it seems useful to repeat in concisely how these answers have been given and, more interesting, why the responses have been done as they have. Following this principle, we will emphasize various facts that seem significant.

Our first purpose is concerned with the theoretical approach. Out of the framework traced in the preface, we would like to stress the importance of mathematical and numerical treatment. This method of assessment is not magic, but by using the experimental data obtained in a particular case, it enables one to make extrapolations toward more general cases. For instance, when the pharmacokinetic parameters have been obtained (e.g., from intravenous administration), the plasma drug profiles can be evaluated with the same drug taken orally whatever the dosage form may be, either immediate or controlled release, provided that the pharmaceutics is linear. Moreover, calculation provides very useful shortcuts, reducing the number of tedious experiments: For instance, when the process is controlled by diffusion, the dimensionless number $\frac{Dt}{L^2}$ clearly shows that the time necessary for a given amount of drug to be released is inversely proportional to the diffusivity D and, more important perhaps, proportional to the square of the lower dimension L of the dosage form.

Another concern relates to the difference between mathematical and numerical treatments: Very often—in most complex cases, in fact—mathematical treatment is not feasible. In contrast, numerical treatment, based on numerical analysis done with the right assumptions, is capable of resolving all the problems. Thus, the mathematical treatment has led to equations in the simple case of intravenous drug delivery, as shown in Chapter 2 and also in Chapter 3 and Chapter 4, with the determination of the kinetics of drug release from oral dosage forms with immediate or sustained drug release. In these chapters, the equations are established in a didactic way, in consideration of the students who will ponder them. On the other hand, numerical treatment is required to solve the problems when an analytical solution cannot be found. Instead of equations, numerical treatment leads to a solution, often called a *model*, which is correct if we think of the assumptions on which this theoretical part is built—these assumptions give a clear idea of the process. It is also called a numerical *program*, meaning that this program can be used on a computer by anyone, provided that he or she owns it on a diskette or a CD. Various examples can be given where the numerical treatment has been necessary: drug transfer through the tissues because of the moving drug concentration in the blood in Chapter 9; transdermal drug delivery in Chapter 10; and also the plasma drug profiles with sustained-release dosage forms in Chapter 6 through Chapter 8.

The results may be expressed in terms of equations or numerical programs, as stated already, but also by curves drawn in figures. Only a mathematician could perceive the reality of a process by looking at an equation or a program. But from the first glance, it is obvious that the curves drawn in a figure provide a clear understanding of the result. For instance, in Chapter 2 regarding intravenous drug delivery, the medical practitioner can deduce the correct way to vary the dose used in the therapy from studying the figures. In Chapter 8, the pharmacist can demonstrate to a patient the importance of compliance by simply displaying the plasma drug profile drawn in various cases of noncompliance. The efficient effect of increasing the permeability of the skin (stratum corneum) appears clearly in the figures in Chapter 10.

Special attention has been focused on patients' intervariability. From the data shown in the literature, it appears that the main pharmacokinetic parameters vary greatly from one patient to another (e.g., the apparent plasmatic volume by 300%, and the rate constant of elimination [or half-life time] by 400%). Thus in Chapter 2, devoted to intravenous drug delivery (which is the basic therapy practiced at the hospital), some methods have been developed to help medical practitioners evaluate the typical pharmacokinetic parameters of the patient from the first bolus injection or infusion that are necessary to adapt afterwards the treatment to the patients.

Two main types of oral dosage forms with sustained release are on the market, depending on the nature of the polymers through which the drug is dispersed: one controlled by diffusion, in which the polymers are stable and pass through the body unchanged, the other controlled by erosion (not mentioning those that swell to such an extent that their kinetics of release look like those obtained with erosion). Dosage forms able to sustain the drug over a long period of time are of great concern, but they need either floating systems (in the case of diffusion) or systems adhering to the GI wall (in the case of bioerodible polymers).

Emphasis has also been placed on the limited reliability of patients. The patient's compliance is worse than is often believed, and Chapter 8 is devoted to how various kinds of noncompliance affect bioavailability. The drug profiles clearly show that, for instance, the effect of an omission on one day cannot be corrected by a double dosing the following day.

Master curves have been built using dimensionless numbers in various cases, as much as possible (e.g., in Chapter 2 with intravenous delivery, but also in Chapter 3, Chapter 4, Chapter 6, Chapter 7, and Chapter 8 with oral dosage forms). Such curves are very useful, as shown in Chapter 2, when instead of time, a dimensionless time expressed in terms of the half-life of the drug is used, leading to a single curve that describes the bioavailability obtained with repeated doses whatever the drugs, provided that their pharmaceutics are linear. In the same way, with oral dosage forms, the amount of drug released up to time t as a fraction of the amount of drug initially in the dosage form brings some advantages. Nevertheless, after using these dimensionless numbers, conventional calculation has been made with the evaluation of the actual concentration of the drug in the blood. It is also worth noting that the diagrams expressing the half-life time of the drug itself (obtained through intravenous delivery) vs. the half-life time obtained with oral dosage forms, whose release is controlled either by diffusion (Chapter 6) or by erosion (Chapter 7), should be of help to readers working in the pharmaceutical industry, because these curves provide the characteristics of these dosage forms necessary to obtain the desired kinetics of drug release and thus the desirable plasma drug profile.

Chapter 5 provides a bird's-eye view of the bibliography on in vitro–in vivo correlations. An attempt has been made to show how useful these correlations are for immediate-release dosage forms, proving that the dissolution phenomena could be more complex along the GI tract than in the dissolution cells. Moreover, it has been pointed out that at time 0, because there is no drug released from the system and no drug absorbed through the GI membrane, the curve expressing the correlation passes through the point 0. Thus there is no way to get a straight line representing in vitro–in vivo correlations, but a curve with a parabolic tendency at the beginning and tending to a straight line fits these correlations better. Of course, this parabolic–straight line mixed curve does not represent the actual phenomena that take place in both in vivo and in vitro systems. A correct equation would be obtained with the series expressing the amount of drug transported by diffusion (Equation 10.5) leading to a curve similar to that depicted in Figure 10.3; thus the straight line could be observed only when the stationary state is attained, after a while when the processes in these two systems are driven under transient conditions.

Finally, a few words about the data collected in tables. Only a few tables are given when necessary, for instance, in order to display the various pharmacokinetic parameters of the drugs considered in the studies, and also in Chapter 2 to provide the precise values of the concentrations at the peaks and troughs in a way complementary to the curves.

Author Index

Subject Index